Nature, Risk and Responsibility

Discourses of Biotechnology

Edited by

Patrick O'Mahony
Director
Centre for European Social Research
University College
Cork

Consultant Editor: Jo Campling

Routledge
New York

Published in 1999 by
ROUTLEDGE
29 West 35th Street
New York, NY 10001

Library of Congress Cataloging-in-Publication Data
Nature, risk, and responsibility : discourses of biotechnology /
    Patrick O'Mahony, ed.
        p.   cm.
    Includes bibliographical references and index.
    ISBN 0–415–92290–9 (hbk.). — ISBN 0–415–92291–7 (pbk.)
    1. Biotechnology—Social aspects.   2. Biotechnology—Moral and
ethical aspects.   I. O'Mahony, Patrick, 1957–   .
TP248.23.N38   1999
174'.96606—dc21   98–37245
                                                                         CIP

This book is printed on paper suitable for recycling and
made from fully managed and sustained forest sources.

Printed in Great Britain

Contents

Acknowledgements

The author would like to thank the European Commission, Director General XII, for their generous support in putting together this collection of essays and in particular Mr Andreas Klepsch. The collection, however, in no way reflects the views of the Commission which bears no responsibility for its content. I would also like to thank the Centre for European Social Research and its staff, especially Séamus O'Tuama and Shannon Whyte, for providing invaluable expertise, time and resources. The contribution of two of the authors, Tracey Skillington and Orla McDonnell, for advice and comments is also gratefully acknowledged, as is that of Jo Campling. Finally, I wish to recognise the authors of the various pieces in this collection for patience and perseverance throughout the long and complicated process of producing a coherent collection.

Notes on the Contributors

Barry Barnes is Professor of Sociology at Exeter University. He has written extensively on science, technology and expertise, and on social theory. He is currently preparing a book on choice, agency and responsibility which, amongst other things, considers the relationship of voluntaristic discourse and the causal discourse that is increasingly prevailing in accounts of human behaviour in the wake of a proliferating biotechnology.

Alfons Bora is a senior social researcher at the Wissenschatzentrum, Berlin. His main fields of interest are sociological theory, sociology of law and sociology of science and technology, subjects on which he has published widely.

Christian Byk is Legal Adviser in the International Law Office, Ministry of Justice, Paris, Vice-President CIOMS, a member of the Human Genome Organisation and former adviser on bioethics to the Secretary General of the Council of Europe. He is the author of many books and articles on bioethical issues.

Gerard Delanty is Senior Lecturer in Sociology at the University of Liverpool. His research interests include ethics, collective identity and methodology, on which he has written extensively. He is editor of the *European Journal of Social Theory*.

Marion Dreyer is academic assistant at the Institute for Sociology, University of Munich. Her research interests and publications focus on public communication in biotechnology.

Orla McDonnell is a lecturer in the Department of Sociology, University College, Cork. Her current interests are in the sociology of the body, health and medicine, and feminist studies.

Ruth McNally is ESRC Research Fellow on forensic DNA profiling in the Department of Human Sciences, Brunel University. He research interest is the social relations of genetic engineering, on which she and Peter Wheale have published three books and many articles and constancy reports.

John Murphy is a lecturer in the Faculty of Law and the Centre for Social Ethics and Policy at the University of Manchester. In 1997 he was appointed fellow of Liverpool and Manchester Universities' Institute of Medicine, Law and Bioethics. His recent publications are on health care and confidentiality.

Patrick O'Mahony is Director of the Centre for European Social Research, University College, Cork. He has written a wide range of articles and reports on the sociology and politics of nature, and has led several large, cross-national research projects on these topics.

Tracey Skillington is a senior researcher at the Centre for European Social Research, University College, Cork. Her major research interests are in political communication on the environment as well as discourse and social change, on which she has published extensively.

Piet Strydom is statutory lecturer in sociology at University College, Cork. His interests include the philosophy of the social sciences and social theory, particularly critical theory, on which he has published extensively. He has directed research at the Centre for European Social Research, University College, Cork, on sociocultural and institutional change, social movements and environmentalism.

Peter Wheale is a lecturer in the Surrey European Management School, University of Surrey, and a director of Bio-Information (International) Ltd. He has contributed many articles on economics and the social and legal aspects of science and technology, and reports appraising the ethics of modern biotechnology.

Introduction: Biotechnology, Uncertainty and Contestation

Patrick O'Mahony

UNCERTAINTY AND RISK PERCEPTION

Biotechnology, the capacity to modify or invent plant and animal life genetically, exhibits two characteristics of late twentieth-century information and communication technologies, one it shares with these technologies in general and in the other it belongs to a more restricted club. In the first case, biotechnology, like other new technologies, involves such far-reaching implications for the organisation of social life that there is little precedent by which to judge its potential impact. What Weingart (1978) has called the orientation complex of technologies – the social, cultural, political and economic forces that shape the trajectory of their development – is, in the case of biotechnology, even more than the new technologies in general, uncertain and contradictory. This lack of clarity derives from the massive implications of the full-scale institutionalisation of biotechnologies, which could foreseeably result in the unbinding of many of the values and rules that currently constitute nature, life and social organisation. It also derives from the fact that these are, to a greater or lesser degree, contested technologies. Given the scale of the potential implications, the future horizons of biotechnologies are shrouded in obscurity, where optimism and anxiety mingle.

In the second, more restricted, shared characteristic, biotechnology helps to undermine the assumption that technologies are neutral: that it is only in their applications that undesirable consequences may ensue. Certain technologies of mass destruction, and even 'civil' technologies such as the automobile whose polluting side-effects are only now being fully appreciated, share with biotechnology the incalculability of accurately assessing the damage they may do. The way in which it is supposed they will influence natural or life processes at the time of their use may not be the way in which they in fact do so. The latter may be far more harmful than anticipated or even possible to know, and are intrinsic to the technology. This is over and beyond the other aspect: that technologies depend for their responsible use on humans who cannot be assumed to act responsibly

1

in all cases. Humans have a poor record in responsible action in the twentieth century and this makes it difficult to deny with conviction that the existence of certain risky potentials will not, at least to some degree, result in damaging consequences.

Observations on the uncertainty of the implications do not mean that it is a proven case that all biotechnological innovations are a bad thing and should be stopped. It is at this point merely to document why biotechnologies are, speaking generally, mired in controversy. Adversaries in the biotechnology debate often only agree that there are profound disagreements. However, in some application areas differences are less pronounced than in others. Plant biotechnology is on the whole less controversial than animal biotechnology, which in turn is much less controversial than human biotechnology, as the deep anxiety over the potential to clone human beings has demonstrated. A range of differences are present even within the same branch of biotechnology. In medical applications in Germany, for example, there is a more or less positive attitude to somatic cell gene therapy which corrects genetic deficiencies, while there is a negative consensus against germ-line gene therapy which removes genetic deficiencies transmitted to offsprings, and an even more negative consensus against eugenic engineering to improve complex human traits such as intelligence (Wessels, 1994; see also Beck, 1995a). The overall impression generated by the diversity of the situations produced from the reactions between technical, natural and social forces is that of a field of great complexity and conflict potential.

BIOTECHNOLOGY INNOVATION AND THE POLITICS OF NORM BUILDING

Biotechnological innovation shares many of the characteristic features of the current innovation climate. First, as a high-tech innovation perceived to have enormous commercial benefits it is actively encouraged by national governments, and in Europe by supranational institutions. Considerable resources have been expended to pump-prime scientific-technical innovation, and this is currently beginning to yield sufficient indications of commercial potential to attract investment by small and large companies alike.[1] Second, it is a reflexive technology in that it allows for constant alterations to techniques and processes depending on experiences encountered in the application environment. In this way, it breaks down barriers between basic research and application, science and technology. Third, as already observed, its implications are uncertain in a general climate in which the process of social embedding of new technologies in social, cultural and political environments is highly unpredictable and contradictory, unlike the situation facing new industrial technologies of the recent past, where political, social and cultural structures were comparatively stable. Biotechnology is entering an environment in which structures and identities are being reconstituted and in which the interaction of technology with the natural and the social is characterised by multi-sided intransparency, not least manifested in the increase of the sense of the riskiness of certain kinds of technological development, of which biotechnology itself is a prominent example.

The first two aspects of the innovation climate – high official legitimation together with good economic prospects and the reflexive form of the technology – are important conditions influencing the third – risk construction – which is the main interest of this volume. Official legitimation and increasing economic viability are the driving force of a rapidly evolving technology whose downstream implications are uncertain in a climate of general political disorientation over how to harness the potential of new technologies. The interplay of rapid technological innovation and the shaping influences present in the wider socio-cultural and political environment in the 1980s and 1990s has given considerable impetus to a neo-liberal ideology as an action-orienting code. Neo-liberalism has sought to promote values and behaviours associated with market freedom achieved through deregulation and privatisation, as opposed to state interventionism, and to promote a climate in which economic interests rather than societal value consensus is the predominant factor in shaping the direction of social change. Technological change since the 1980s has therefore taken place in a climate where freedom of innovation and the 'inevitability' of scientific-technical progress are stressed and where issues of legitimacy and fairness have been relatively marginalised. Biotechnology has occupied an ambivalent relation to this process. On the one hand, as a technology whose risks have at some level to be acknowledged even by its most ardent proponents it has been more than ordinarily circumscribed by value barriers to its further development. On the other hand, given the impossibility of realising value consensus on certain research paths as they currently stand, it has proceeded towards innovation, taking advantage of the neo-liberal climate, without wide-ranging societal legitimacy. As von Schomberg (1995) observes, biotechnological developments without controls produce *faits accomplis*.

Amongst the general public, biotechnologies are perceived – along with other technologies such as nuclear technologies, certain pharmaceutical processes and many other industrial and energy technologies –

as intrinsically hazardous. Both the factual perspective of the actual or probable creation of hazards and the evaluative standpoint that hazards are possible (though hard evidence cannot be found to prove this) compel what might be described as constraining regulatory action, such as qualifying the freedom of certain kinds of research or prohibiting certain other kinds altogether.[2] By contrast, if the full range of biotechnological potentials are to be realised in most or all of its sectoral areas, positive regulatory action is needed in the form of appropriate juridification to create property rights, and balance them with other human and non-human rights. A routine legal and administrative basis for bureaucratic action also needs to be created.[3] Necessary conditions for the adequate juridification and general governance of biotechnology are currently absent on a number of counts. First, the prevailing uncertainty with regard to the scientific assessment of causality and prediction of consequences leaves legal and regulatory knowledge seriously deficient and unable to provide citizens with recourse to the law when their rights are infringed, or provide sanctions for the infringement of legal norms (von Schomberg, 1995; Bora, this volume). Second, the absence of minimum value consensus leaves both regulatory regimes and the law without sufficient legitimacy, and issues are either left permanently undecided or are decided on an arbitrary basis.

The extreme difficulty of establishing minimal value consensus on biotechnology derives from the fact that it brings into play issues of a deep-rooted ontological nature. It pits an anthropocentric utilitarianism still wedded to the idea of progress through technology against a 'nature-defending' deontological philosophy, which has become suspicious of the virtues of scientific-technical advance.[4] While it is possible in the related field of environmental politics to discern the emergence of at least some value agreement between contending actors in the emergence of a new political ecology (Beck, 1995; Eder, 1996; O'Mahony, 1996) this is predominantly not the case in the politics of biotechnology. In this case, such fundamental disagreement and confusion about the wider implications persists that no 'idealisation' (Habermas, 1990; O'Mahony and Skillington, 1996; O'Mahony, 1996) is apparent that would bring a degree of convergence to divergent positions. Some sense of the difficulty of finding factual and evaluative agreement on biotechnological issues can be gained if it is compared with the most divisive moral issue currently on the political agenda: abortion. While opponents of abortion occupy fundamental deontological positions, abortion is still legal, whereas human cloning or certain kinds of biomedicine are not. The additional problems in the case of biotechnology involve the position of third parties. In the case of abortion there are no long-term health issues other than those affecting the pregnant woman. In the case of biotechnologies, by contrast, individuals, groups and even the species itself can claim a direct physical interest in the safety and visibility of actions taken.

Public anxiety created by the uncertainty and disputability of causes and consequences is compounded by the cultural threat offered by bio-innovation to the anthropocentric sense of 'the natural'. While advanced genetic engineering appears on one level as the ultimate expression of an-thropocentric control over nature, at another level the sense of the perme-ability of all boundaries disturbs the existing sensibility of what it is to be human and the hierarchies by which humans situate themselves amongst other life-forms. Advances in biotechnology have the potential to change the prevailing experience and understanding of being human by, for example, delaying the ageing process, selecting the sex of offspring, con-trolling for intelligence, even creating a new kind of neo-humanity equipped with greater physical attributes or lesser or greater intelligence depending on preference. The options created by biotechnological advance are rapidly bringing humanity to an evolutionary crossroads in which a fundamentally altered species could be made possible. New evolutionary potentials extend beyond the species to its natural and social environment. Transformations in animal and plant life could change the face of agricul-ture. Further, the options afforded by the use of biogenetics in the social services are considerable. Genotype screening, which allows for the more precise identification of categories of risk, fit in with a general political cultural pattern of the individualising of collective securities, a process which is eroding the commitment to mass public provision in western democracies (Beck, 1995; Skillington, this volume). Skillington (this volume) draws attention to the manner in which the application of biotech-nological advances in medicine and the social services leads to an empha-sis on ascriptive differences between bodies rather than a common affinity with nature. Recreating the substratal text of nature that has hitherto operated as a common, underlying topos for human self-interpretation (Harraway, 1995) has profound implications for ideas of identity and their application, and adds to the underlying difficulties of establishing common values.

DISCOURSE AND THE INSTITUTIONALISATON OF BIOTECHNOLOGY

Given the scale of the implications, it is not surprising that biotechnolo-gical innovation has not been followed by consensus on legal and practical norms. In the chapters that follow both value conflict and prospects for consensus are central motifs. For the purposes of examining the existence of dissensus or the prospects of consensus, the analysis of socio-political discourses is fundamental. It is possible, and becoming daily more possible, to clarify the extent of the new ontological possibilities offered by biotechnology. It is also possible to construct scenarios in which even further out-products of bio-science could become materially feasible. What remains deeply uncertain is the intractable cultural issues that are associated with these technologies, which centre not on 'what' and 'how' questions, but on ethically guided 'why' questions. These ethical questions can be identified through the perspectives present in various discourses. Cultural discourses on biotechnology cannot at present evade an acute evaluative and hence political status.

The chapters in Part I, *Theoretical Reflections on Biotechnology and Responsibility*, which deal broadly with the theme of modernity and the discourse of responsibility, explore the wider culture of political-normative disputation, both in general public discourse and in the institutionalised discourse of experts. Responsibility emerges as the central focus of these chapters in two specific senses. First, the creation of qualitatively new kinds of risk, which Beck (1995, p. 78) describes as 'politically explosive hazards which render questionable the principles of calculation and precaution', leads to a new discourse of responsibility towards nature. Second, and relatedly, responsibility is a central element of those experts who are charged in the glare of responsibility-demanding public opinion with deciding under what conditions certain kinds of research, and exploitation of research, should proceed and how to regulate the downstream consequences.

Part I begins with Strydom's 'The Civilisation of the Gene: Biotechnological Risk Framed in the Responsibility Discourse'. Strydom understands risk in terms of the processes that involve the metamorphosis of collective goods into collective bads. Biotechnological risk, because it penetrates into nature in an unprecedented way, is central to the situation of contemporary modernity as a risky state. Together with other processes, such as environmental degradation, biotechnological advances reveal the dependence on a natural substrate which was hidden in industrial modernity. Now, the implications of technical advance are immediately examined within the newly emerged 'constructivist' impulse of the social sciences. The constructivist gaze of the social sciences radicalises the freedom of communication contained in the discourse of modernity by revealing how new ideas, artifacts, systems and organisations depend on communication for their collective production. If the communicative processes involved in constructing states of affairs can be made 'reflexively' available to actors, then a higher degree of complexity can be sustained or voluntarily relinquished as too risky. Constructivism in the social sciences currently reacting to an environment of actual and perceived risk is turning to the exploration of the cultural foundations of the contemporary relationship between humanity and nature, and in this process amplifies the new societal discourse of responsibility. The discourse of responsibility combines the re-orientation of perceptions of nature in relation to the body and the planet as home with the idea of a communication community allowing participation on the basis of equality and fairness. Only the societal realisation of the latter could facilitate the emergence of what is needed: a planetary macro-ethics of responsibilty. Such a macro-ethics must be located constitutively within the metaphorical as well as practical significance of genetic science. This is the significance of Strydom's reference to the civilisation of the gene.

Delanty follows similar ground in situating biotechnology within the framework of responsibility. The focus of Delanty's differs from Strydom's, whose considerations of biopolitics is oriented to broad, contestatory discourses about cultural assumptions and social arrangements. Delanty extends Strydom's focus to consider directly the conditions necessary for a dialogical and rational biopolitics. While biotechnology currently shows the limits of enlightenment rationalism with its assumptions about the pre-social neutrality of all forms of technology and the acceptability of a fallabilistic science, Delanty argues, following Heller and Feher, that opposition to biotechnology cannot retreat into an 'essentialist' fundamentalism, which refuses to budge from its own ontological ground. Delanty's position implies a constructive view of the possibilities of biotechnology which, while critical of any attempts at irresponsible autonomy by science and technology, refuses to follow oppositional movements into outright opposition. This is also the substance of Lenk's (1992) critical appropriation of Jonas (1984). The implication is that how far biotechnology should be allowed to proceed in certain areas cannot be foreclosed, but must be decided by rational deliberation on the substance of responsibility in the light of what is factually known about the risks and reasoned argument from evaluative standpoints.

Barnes' essay completes Part I by considering the relationship between expertise, trust and responsibility. If substantive and procedural responsibility is actually to be achieved in the field of biotechnology, then some form of mediation between lay and expert knowledge will be required. Barnes addresses the problems that arise in achieving such a mediation.

Principally, he returns once more to the issue of the ontological in considering the relationship between technical and evaluative expertise within a sacred/profane framework. The boundary between the sacred and the profane is constantly shifting; biotechnology provides yet another frontier. Barnes holds that while this boundary may shift, the condition of human beings as 'persons/souls, living entities, active agents with responsibility, conceived as individual wholes' (Barnes, this volume) will always be distinguishable from the absorption of the human body into an instrumental framework. This correlatively means that biotechnological expertise can never be strictly technical expertise in so far as it presses on this distinction. If expertise is to have authority in this domain, then this authority will depend on *moral* trust.

The establishment of such trust depends upon fair procedure and open minds. In pursuing these themes Barnes elaborates on issues raised by Strydom and Delanty about the conditions and arguments of oppositional movements. Common to all three writers is the idea that oppositional movements provide as it were a baseline against scientific-technical colonisation of still 'reserved' domains. If they are to be shifted from such ground, then they must be reasonably persuaded by those who pursue further technicisation. Anything else would be ethically unacceptable and risky. This, of course, carries the corollary raised by Delanty that the defence of reserved positions must also be reasoned. Barnes finds that current procedures for deciding in contested domains are weighted too far in favour of the technical and therefore act against those who oppose them. His condition that expertise on biotechnology is legitimately an issue of moral trust means that the moral claims of oppositional movements cannot simply be met with technical arguments. They have to be met with moral arguments too – a factor that is insufficiently taken into account in various procedures, such as science courts or technology assessment procedures, or in the prevalent view that the arguments of oppositional movements are simply 'irrational'. At the same time, these movements have to recognise that in order to contest the autonomy of the technical they have to struggle within its parameters. Hence, movements are developing impressive media skills in presenting their argument in the public domain, where currently in many disputed cases in the absence of effective institutional procedures the issues of moral responsibility and factual evaluations are played out.

Barnes' essay, in its concern with placing responsibility in the context of the interplay of lay and expert cultures in the public sphere, is a bridge between Parts I and II. Part II is entitled *Constructing Values: Public Communication on Biotechnology*. Many biotechnological innovations stand at the point where innovation potential but not public legitimation has been demonstrated. In this phase, media constructions of the implications of these technologies through symbolic exploration and argument are critical. The relative weakness of institutional mechanisms for decision-making in contentious cases that are both effective and legitimate[5] makes this all the more so. The three chapters in Part II are of their nature less general than those in Part I, but do present a good insight into the value of constructivist approaches. The assumption that underlines all three is that as biotechnology transgresses into reserved moral spheres, moral and normative public discourse acquires unusual importance. McDonnell's essay, places public discourse in Ireland on new reproductive technologies within the framework of the international debate; Dreyer examines the quite different debate on the use of biotechnological processes in the German pharmaceutical industry; while O'Mahony and Skillington move on to a more general level, identifying typical 'discourse coalitions' on biotechnology in the British and Irish press.

McDonnell's essay provides an empirical case study which complements Barnes' exploration of the relationship between experts and the public. His essay explores the Irish case where expert medical cultures in association with the Catholic Church have enjoyed a monopoly over ethical rule-making. New reproductive technologies such as in vitro fertilisation disturb this partnership and call for creative action. While experts seek to confine creative rule-making to professional ethics bodies, new 'discourse publics' (Fraser, 1989) seek to broaden it out to reflect, for example, women's experiences. Such broadening is made difficult by the trajectory of international public discourse on the issue. McDonnell characterises this (following van Dyck, 1995) as progressing from establishing a sense of a need for new reproductive options in the 1970s and early 1980s, to the absorption of a feminist counter-myth in the 1980s, to the portrayal of science as the dominant episteme and the construction of legislative frameworks in the late 1980s and early 1990s, and since that time to the dominance of a rights discourse. McDonnell shares with Barnes the contention, consistent with van Dyck's periodisation, that scientific-medical expertise has dominated the framing and practice of applications of the new technologies, although she identifies in the Irish case – again in line with international experience – the emergence of new discourse publics, which aim to subvert existing power relationships. These power relationships are by contrast sustained by such discursive themes as the 'narcissistic woman' which construct women in a market-compatible manner and which serve the dominant epidemiological script. This dominant script can be strategically deployed to prevent subversion by negating potential fields of struggle and resistance. The question at the heart of McDonnell's essay is whether a new discourse public ensuring a central role for the exploration of women's experience can gain influence over the currently dominant scientific-epidemiological script. This discourse public might effectively carry a new ethic of collective care to contrast with the individualism of the current script.

Dreyer's essay explores a case of public construction in which a new discourse public actually has gained substantial leverage. Dreyer examines how the growth in sensitivity to risk amongst the general public, as a generalised phenomenon really only a decade old, has brought new pressures to bear on the pharmaceutical industry, currently the leading user of biotechnological processes, which responds with communication campaigns to defuse and contain the impact of this sensitivity on their operations. Dreyer's essay is similar to McDonnell's in concentrating on an area in which biotechnological processes are currently employed and subject to regulation. Following the basic insight of the neo-institutionalist paradigm, Dreyer claims that the chemical industry in Germany, in the case of firms and peak industry association, increasingly needs to legitimate itself in its societal environment, which is composed of beliefs, norms, conventions, social structures, the political system and organisational fields. It can be described as *embedded* in this societal environment in a way that requires it to adopt at least manifestly a stance of social responsibility. While overt stances of responsibility do not necessarily translate into corresponding behaviour but may only reflect rhetorical strategies, Dreyer concludes that industry, by acknowledging the need to engage in dialogue at all, opens itself up to scrutiny and to public attention and moral pressure.

Dreyer's essay deals with a vital issue of current environmental politics: the growing importance of public communication as a means of regulating complex and changeable processes, especially those like biotechnology associated with high-risk sensitivity (see Hannigan, 1995, O'Mahony, 1996). O'Mahony and Skillington, following this orientation, analytically examine the role of public culture in evaluating risk and defining responsibility. They identify and describe the principal discourses seeking to construct the environment of biotechnological innovation as they appear in the British and Irish press. They use the term 'discourse coalition' to signify a degree of semantic commonality between actors promoting certain messages in the press. Four coalitions are identified. The first of these is the *fundamentalist critique of biotechnology coalition*, which demands new moral standards for regulating human intervention in nature and the prohibition of much that is now considered acceptable, such as experiments on animals. The second is characterised as the *new left liber-*

tarian coalition, which does not carry the anti-science animus present in the first coalition but wishes to see science coupled with regulatory strategies that would guarantee both safety and equity. The third coalition, the *counter-scientific expertise coalition*, is a tendency within science to identify problems with certain kinds of science, opposing sheep cloning, for example, even before Dolly. Finally there is the *biotechnology as solution discourse coalition*, which adopts a rational pro-biotechnology stance, standing on the virtue of societal need and the authority of scientific evidence on risks.

The account of discourse coalitions is illustrative of the path that has to be traversed if responsibility is to emerge from dialogue within a 'communication community' or, put another way, if debate between different positions on the natural, life and social implications of biotechnology is not to be ontologised. Central to this process would be the emergence of a new 'frame' which could to some degree be accepted by all political actors in the field and which would also ensure responsibility over time and across space. In some areas of biotechnology covered in this volume, such as reproduction technologies or the chemical industries, the outlines of such a frame can be discerned in the emergence of dialogue and in regulatory processes, though this is a fragile semantic and material commonality always likely to be disturbed by catastrophe or unprecedented cases. Those areas in which a convergent frame might be possible are where biotechnological innovation is coupled with other themes that have been factually and normatively clarified through preliminary dialogue and regulation. This includes regulation of the pharmaceutical industry on the back of such philosophies as sustainable development and ecological modernisation, and artificial interventions in reproduction such as abortion. However, as McDonnell and Dreyer make clear, even in those areas in which discourse is at some level rationally structured, it remains contentious. Other areas are more controversial, such as animal – and even more so human – cloning, or experiments involving cell tissues. These areas bring out the fullest range of differences and allow for relatively univocal, entrenched positions. Approaches which call for responsibility through dialogue such as those of Delanty or in another way Byk (this volume) have to reckon with this difference.

O'Mahony and Skillington close with an account of the 'opportunity structures' which decide how influential the different coalitions will be. This assessment reveals a differentiated landscape whose future development depends on wider socio-political developments. The fundamentalist coalition has more cultural than institutional power; the new left libertarian coalition may become significant if its message of societally responsible liberalisation gains institutional standing through governing power; the counter-science coalition is becoming increasingly prominent as the full risks and potentials of biotechnological innovation are becoming apparent, creating widespread ambivalence; and the biotechnology as solution coalition moving with the deregulatory political cultural climate is still institutionally favoured. This interplay of cultural and institutional constructions and powers leads into Part III, which deals with issues of 'juridifying', or providing a legitimate and efficient legal framework for biotechnology. Part III, *The Dynamics of Institutionalisation: The Regulation and Juridification of Biotechnology*, concentrates on attempts to provide a legal and regulatory framework in circumstances of ongoing societal argumentation. Where public controversy exists, law becomes the critical testing ground of institutional viability. Law can be distinguished from technology as a secondary medium for organising the social. A new technological systems is, by contrast, a primary medium whose implications require juridification on two counts. First, law must regulate the consequences of innovation so that privately acting agents can pursue their interests; and second, law must defend both the welfare of these agents and the general common good. Law, therefore, is an indispensable medium in the process of institutionalising an innovation. However, before comprehensive legal institutionalisation is possible, a demonstration of material feasibility and a symbolic process of alignment between divergent positions are required. The chapters in Part III explore in one fashion or another the complexities of juridification in conditions where such alignment is either absent or is highly qualified.

Part III begins with Murphy's essay on the legal regulation of assisted conception in Britain. Murphy's consideration begins in a sense where McDonnell leaves off: the situation where medical potential has been demonstrated and a discourse of rights established. This level of advancement meets the preliminary conditions of material feasibility and minimal symbolic alignment to allow juridification to proceed. The law encounters a host of problems for which solutions must be found. These include genetic and social rights to parenthood, confidentiality issues, civil liability and mishap in the case of assisted conception. From a legal standpoint, notwithstanding these complexities, a good case can be made that the right to avail oneself of assisted conception is a fundamental human right, founded in Article 12 of the European Convention on Human Rights and Fundamental Freedoms, which provides a legal right to all citizens of marriageable age to marry and found a family. Regulatory intervention in the case of reproductive technology may, however, be justified on several counts, viz. prohibitions on certain kinds of scientific experiment, safety considerations and conflicts of rights. The essay documents how the law strives – and in the view of the author with a degree of success – to decide on such issues.

Though in vitro fertilisation and other reproductive technologies address very complex issues which derive from unprecedented technical potentials and the contemporary volatility of cultural attitudes, they nonetheless admit of a practice and a regulatory framework which, as Murphy points out, is going to continue whatever opinions may be held on its desirability. They therefore belong, at least in so far as they are currently advanced, to the category of complex but not intractable issues deriving from biotechnological innovation. However, a direct correlation may be traced between the real complexity of the issue and governance innovation in the English regulatory system. The area is regulated through the Human Fertilisation and Embryology Act 1990, which established a regulatory Authority of the same name. This regulatory agency is entrusted by Parliament with the regulation of assisted conception and embryo research and is responsible for issuing licences and establishing rules of practice. The members of the Authority who are appointed by the Secretary of State, are scientists, medical personnel, lawyers, sociologists and clerics. The Authority is part of an innnovation in the style of governance that has taken place in Western Europe in the last decade. This is generated by the state withdrawing from attempts to directly 'steer' certain areas and leaving such activity to societal 'self-government', of which the Authority is an example.

Murphy identifies a number of potential problems arising from regulation being entrusted to such an Authority. First, it means that the British Parliament is not directly responsible for authorising projects and experiments that push at the very frontiers of ethical acceptability. The corresponding legitimation deficit is compounded by the way in which the members of the Authority are appointed by the Secretary of State. Second (a point similar to Barnes') the composition and procedures of the Authority are likely to favour a form of deliberation in which technical arguments put forward by scientists and medics predominate. Third, the authority lacks an executive to police its decisions. Nonetheless, Murphy does regard the Authority as a step forward in depoliticising medico-moral questions and entrusting rule-making to those professionally concerned with assisted conception and embryo research. In this respect, his orientation differs from McDonnell's. McDonnell is concerned with the domination of the experiential cultures of women by rationalised expert cultures, though it could be argued on the basis of Murphy's own work that this difference would be reduced if the Authority were to be made more representative, suitable for the articulation of non-technical arguments, and with an executive arm with clout.

The issues of participation and institutional innovation continue in Bora's essay, where it is the central concern. Bora notes that participation is called for when alternatives for exerting control and influence appear to fall short. This applies especially to decisions on the introduction of new technologies whose social and environmental implications are hard to predict. He claims that attempts to allay the concerns of citizens by involving them in decision-making have not lived up to the positive expectations accorded to them. This applies especially in Bora's account of the legal system. Participation through public hearings or through mediation mechanisms cannot compensate for the manner in which the legal system is left isolated with intractable problems owing to the lack of clear inputs from other systems, such as the scientific and political system. The legal system is therefore burdened with additional responsibility for assessing risk in disputed cases, a burden it cannot discharge and which is exacerbated rather than alleviated by citizen participation. The problem with citizen participation in Bora's empirically schooled judgement is that the kind of arguments that are used would, if accepted by the legal system, result in its de-differentiation and domination by other social systems.

These arguments are what Bora styles substitutive and methodological-evaluative. The first is when another system – business, economics or politics – attempts to impose its code on the legal system, and the second is when the legal system is encouraged to depart from its normal evaluative methods, i.e. through accepting the idea of hypothetical risk rather than probable damage as a criterion.

Bora aptly describes a situation in which judicial processes have difficulty operating in circumstances where values carried in competing discourses are incommensurable. This produces a situation similar to what Byk (this volume) describes as the conservatism of the law. But the problem which Byk treats explicitly and Bora implicitly is also similar: how to create a legal framework in conditions of social and political dissensus? In a sense Bora's essay stops at this point. He speaks of the trend for the law to pass on difficult decisions to legislative bodies. The difficulty of procuring legitimacy for decisions in contested cases and the recognition that resolving these cases will require structures that result in the interpenetration of system boundaries lead precisely to calls, following Willke (1992; see also Giegel, 1992) here, for the strengthening of 'symptomatic action systems' such as round-tables, concerted actions, mediation and others. Bora's essay, while a valuable account of why law becomes over-burdened with risk-related decision-making, still leaves open the question of how consensus should be achieved in such decision areas, or indeed whether the need for at least far-reaching principled consensus should be ignored.

Byk inclines to the latter view. In view of the difficulties of winning agreement on common values, problematic at the national level and even more so at the supranational level, he proposes that the attempt to derive harmonising rules from fundamental legal texts on values should be abandoned. Rather, these rules could be produced from pragmatic consideration and legal precedent much as in the case documented by Murphy. Byk equates the search for common values as being synonymous with citizen participation and he is of the view that such participation could be inimical to reaching possible decisions. This judgement is in part based on the process Byk characterises as the 'deconstruction' of the legal system. The legal system today has to reckon with the rise of a culture of professional expertise whose rationality has penetrated into legal rationality, with a profusion of interests demanding rights of consultation and participation and with a 'thicker' context of advisory and policy-making bodies and a richer spectrum of instruments for the enforcement of rules. Expressed in the system/environmental language used by Bora, the legal system is forced to absorb into its inner complexity the greater contingency arising from its changing environment. Byk responds to this deconstruction by proposing a quasi-judicial process to establish agreement on rules even where common values are absent. But it is clear from his essay that there are no easy answers to questions such as the supranational harmonisation of rules where different societal values and legislative frameworks are present in national contexts or where not just the absence of value consensus but profound value conflict is at issue.

One of the principal areas of value conflict deriving from the implications of biotechnological innovation is the issue of patenting life-forms. Patenting is a legal mechanism to protect the invention and commercialisation of products and processes. In the last essay in Part III, McNally and Wheale show how patenting is intrinsically linked to biotechnological advance. Innovators actually seek out instances of natural biodiversity as inputs into genetic engineering process especially in the case of plant biotechnology in the less developed world, which the authors describe as gene-rich and resource-poor. Patenting plant, animal or human components of life-forms is highly controversial on a number counts. Along with the fundamental ethical objections and uncertain risk assessments already discussed, issues of equity also arise which are central to the concerns of this essay. Biotechnology as a regime of accumulation and regulation

offers prospects of a new pattern of development that will shape economic institutions and social relations well into the future. However, this
developmental model as it currently being juridified in Europe following
the US model offers a 'dystopian' prospect for the institutional order and
for the prospects of the developing countries. Somewhat contrary to the
tone of the earlier essays, McNally and Wheale document how social
mobilisation was translated into political power to block the passing of
the first draft of the European Patent Directive. Their analysis therefore
reaches substantially different conclusions from those of Bora and Byk,
in that McNally and Wheale perceive a deficiency of normative rationality in existing institutions which *must be* corrected by empowering social
mobilisations and bringing them to the heart of institutional decision-
making.

The adequacy of existing rationality is also the subject of Skillington's
concluding reflections. Biotechnological controversies do not lend themselves easily to conclusions and the different perspectives present in this
volume, and the fundamental societal oppositions, are not going to disappear rapidly. Skillington works from a position, which is at least a
minimal common observation of the collection, that the rate of social,
natural and personal change made possible by actual and still more future
biotechnological innovations requires greater collective reflection than has
so far been accorded to it. For her, the inadequacy of reflection cannot do
justice to, first, the too great reflexivity between scientific research and the
creation and satisfying of new 'convenience' needs in conditions of risk
and, second, the actual and potential implications of biotechnological innovations for surveillance, the physical and metaphorical reconstruction of
the body, the extension of inequality, and ultimately, the exploration of a
'post-human' condition in processes such as human cloning or the slowing
of the ageing process. These processes are accompanied by a spiral of
public anxiety which, due to deficiencies of information and opportunities
for deliberation, has not progressed towards a sense of control. According
to Skillington, control is moving elsewhere in the civilisation of the gene
(to use Strydom's term). It is moving towards a mode of governance based
on bio-control, a process marked more by the technical power of corporations to control life-processes and associated social organisation, rather
than by models of institution-restraining citizenship. Skillington's essay,
while recognising the need for collective responsibility, points to the
difficulties of achieving it in cultural circumstances where collective solid-
arities and forms of organisation are under threat. In the case of biotechnology this becomes paradoxical in that while biotechnology demands
collective responsibility, it also provides means to undermine it further

through new possibilities of surveillance and individualistically tailored collective provision.

The volume as a whole is concerned with exploring how the contested domain of biotechnological innovation has far-reaching implications for how concepts and practices of responsible governance can evolve. The social theoretical diagnosis offered by many of the contributors point to the problem – a responsibility deficit in the face of risk – and the space of the solution – some variation on improved communicative governance – but cannot do more than offer tentative suggestions as to how effective and legitimate governance could be provided. This reserve can be explained as a product of several factors. First, it reflects the orientation of social theory not to extend itself beyond characterising desirable normative orientations. Second, as Bora points out, exploring institutional innovation risks the reversal of existing differentiations and interpenetrations between social systems in circumstances where there is as yet insufficient evidence as to how direct, participative democracy can effectively function within the ensemble of democratic institutions. Third, many of the more intractable issues in biotechnology could not be communicatively resolved given the current distance between contending parties. Put another way, a governance semantics is absent, and in its absence governance innovation cannot be expected. This leaves many complicated ethical issues somewhere in the interstices between irreconcilable perspectives in the public domain and limited institutional capacity to act as a consequence of inadequate deliberative and decision-making capacity. Deciding on many of the issues associated with biotechnology leaves few easy options and in some cases no options at all within current ethical sensibilities and institutional regimes. Of one thing we may be certain. Before certain frontiers can be crossed or definitively not crossed, considerably more cultural adjustment, deliberation and procedural innovation will be required.

Part I

Theoretical Reflections on Biotechnology and Responsibility

1 The Civilisation of the Gene: Biotechnological Risk Framed in the Responsibility Discourse[1]

Piet Strydom

> ... one of the fundamental tasks of social science, if it is not to be stupidly positivist, is to think the structures of thought ...
>
> Pierre Bourdieu

> ... eine diskursive Organisation der kollektiven Verantwortung für kollektive Handlungen auf nationaler und internationaler Ebene...
>
> Karl-Otto Apel

INTRODUCTION

Ours, at the end of the twentieth century, is the civilisation of the gene. The gene is the leitmotif of our era. It is not only the natural and life sciences that bring this home to us, but the social sciences too. The biological and life sciences are in the ascendant at the expense of the physical and mathematical sciences, which formerly occupied the central position. In the 1960s as one era drew to a close and a new one unfolded, the technology of the civilisation of the atom or inert matter was no longer interpreted simply as an efficient and cost-effective means of overcoming temporary problems in harmony with social advancement. Rather it was seen as raising the issue of environmental degradation and the possibility of the disappearance of all life on earth. Not long after, the revolution in molecular biology and biotechnology, marking as it did the arrival of the civilisation of the gene, began to demonstrate not merely that we have acquired an exciting or, rather, frighteningly extensive and intensive capacity to intervene in nature, but that we are in fact able to constitute nature, indeed, to create our own human nature ourselves.

Amidst a persistent widespread perception of a crisis in the social sciences since the 1960s, related to this transition from the metaphor of the atom to that of the gene, they have in the last two decades finally come into their own, as both disciplines and discursive practices. Everything is seen from the social scientific perspective, which itself is no longer confined to the expert culture of social scientists but has penetrated society itself. Everything is seen as the outcome of social construction. All creativity and innovation, every attempt to furnish the world with new ideas, standards, objects, artifacts and so forth, involves social processes – protracted processes of co-operation and competition, struggle and conflict, accomplishment and failure. From gender and sexuality through knowledge, technological artifacts and science to nature itself, everything is seen as socially constructed. The basic point, made many years ago by Moscovici in his seminal work *Essai sur l'histoire humaine de la nature* (1968) when he raised the fundamental question of our time, the question of nature together with culture, subsequently elaborated upon by such authors as Ulrich Beck (1992) and Klaus Eder (1988), has recently been restated by Irving Velody (1994, p. 83) in a wonderfully pointed – even if one-sided – way: 'the social construction of everything signals the final success of social science: the expulsion of nature and its replacement by culture.' It is remarkable, furthermore, that the constructivism of the social scientists, with which they seek to grasp the constructive activities of social agents, makes central use of concepts analogous to the gene or genetic code. Among them are such concepts as the 'symbolic code', 'cultural code' and 'binary code', to mention only the most prominent sociological examples.

The gene as the leitmotif of our era, to which biotechnology is absolutely central, finds its most pressing manifestation, however, neither in the natural nor in the social sciences, but rather in the fundamental question of our time – Moscovici's (1990, p. 7) 'question of nature'. What for Moscovici is the primary question of our time is for Habermas (1992, pp. 196–201) a collective problem around which a discourse, consisting of different discourse types and forms of negotiation, takes shape. The primary question is thus best approached through communication and discourse. Since the breakdown of the religious-metaphysical world-view and the monopoly of the clergy over the public interpretation of the world, since the emergence of a diversified intelligentsia and the penetration of social relations by communication in the early modern period, this has been the characteristic feature of society. From the outset, modernity has been a communicative and discursive phenomenon. At the end of the twentieth century, however, the civilisation of the gene has rendered it doubly reflexive. It is no longer simply a matter of the collectivity addressing the problem of the constitution and organisation of society through the question 'What should we do?', but it is one of self-constitution and self-organisation in terms of self-constituted symbolic or cultural codes through the question 'What should the human species do?'

In this chapter, I start from suggestions deriving from Moscovici and Habermas, Apel and Bourdieu, in an attempt to locate the debates about biotechnology in the context of the contemporary responsibility discourse, as I propose to call it, which in turn belongs within the more general context of the discourse of modernity. The aim of this discourse-theoretical contextualisation of biotechnology is to reflect on the discursively activated cultural conditions of the civilisation of the gene. This perspective affords a glimpse of the structuring of the contemporary discourse, and hence suggests explanations for the formation of social identities and the deconstruction of existing and the construction of new social groups or collective agents. It specifically contributes to the clarification of the cultural forms and social and political practices that are currently shaping the possibility and modalities of a more discursive resolution of collective problems, a more pluralist participatory politics, and corresponding reflexive institutional innovations. As regards biotechnology, more specifically, the perspective adopted here allows one to begin to appreciate how the dramatically increasing control of human nature itself can be controlled and, indeed, is being made amenable to control – even if of an indirect kind.

To establish a point of contact, a brief description of biotechnology is ventured in the following section. This is followed by two further sections. The first is devoted to the discourse of modernity, while the other characterises the current responsibility discourse through a comparison and contrast with preceding historical discourses. The principal thrust of the argument is the twofold thesis (i) that biotechnology is best understood as forming part of the current collective problem of risk, which is addressed through the question of nature within the context of the responsibility discourse; and (ii) that all this should be seen as occurring within the field opened up by the discourse of modernity. The implication is that the debates, controversies and conflicts around biotechnology, far from being isolated or confined, are central to the new and unique way in which the characteristic modern problem of the self-constitution and self-organisation of society is being resolved in the late twentieth century for the twenty-first century.

THE BIOTECHNOLOGY REVOLUTION

Since it is based on one of the leading forms of scientific knowledge production, biotechnology occupies a special position among the so-called 'high technologies'. In its current understanding it is a generic term referring to a variety of scientific practices and techniques which derive from the contributions of theoretical molecular biologists and are employed for the commercial exploitation of biological organisms and processes. In plant breeding, it takes the form of tissue cultures and genetic engineering and manipulation aimed at improving crop disease- and drought-resistance, eliminating the need for nitrogen fertilisers, and increasing yield and protein levels. In the industrial context, it finds application in particular in the medical field involving, for instance, large-scale fermentation processes aimed at the production of naturally occurring proteins, monoclonal antibodies and the development of human growth hormone. Most central to both science and industry, however, is the mapping and sequencing of the full chromosomal structure of plants, animals and humans which will allow not only DNA probes to pinpoint genetic disease, but also biotechnological inventions (e.g. genetically engineered or so-called 'transgenic' organisms).

While the innovations on which developments such as these depend mostly date from the 1980s, the principal scientific breakthroughs occurred in the 1940s, 1950s and 1970s in the course of protracted theoretical research. The first key achievement to provide the basis of the biotechnology revolution of the 1970s and 1980s was the identification of bacterial plasmids, which are crucial for the manipulation of genes and the transfer of genetic information from cell to cell. This was followed by Watson and Crick's description in 1953 of the double-helix structure of DNA (deoxyribonucleic acid), the biological polymer forming the genetic material of all living organisms. The most important landmark on the eve of the biotechnology revolution, however, was the identification in the early 1970s of the catalyst (i.e. restriction and ligase enzymes) which allows the dissecting and rejoining of DNA and hence the isolation of individual genes – a development known as recombinant DNA. On the one hand, it is the virtually unlimited potential projected by these scientific achievements for the reconstruction of nature through genetic engineering that accounts for the intense interest in biotechnology on the part of pure and applied science, industry and the state. On the other, it is the articulation of such scientific and technological innovations with political and legal developments (e.g. the patenting of biotechnological inventions) and the globalisation of free trade and intellectual property rights in biotechnological inventions that provides the necessary conditions for the marked increase in controversies and conflicts involving the public and the new social movements.

Academics acting as small-scale entrepreneurs had been the key players in the initial commercialisation of biotechnology (Street, 1992, p. 86). In the wake of these embryonic biotechnology industries, large corporations such as Eli Lilly, Genentech, Monsanto, Shell and ICI were among the first to move into the biotechnology field (Webster, 1990, p. 199). Although considerable commercial investment has not yet been justified by the availability of a corresponding number of successful commercial products, corporations remain intensely interested in the control of plant pathology without resort to chemical treatment (Webster, 1990, p. 185) and in the genetic correction of defective inherited molecules, such as those identified in the case, for example, of sickle cell anaemia (Street, 1992, p. 106). As regards state interest, an OECD report (OECD, 1988; Kreibich, 1986, pp. 418–57) found a wide range of variation among the 15 countries surveyed, both in terms of arrangements established and the spread of biotechnological applications covered. Yet at the same time it makes clear how countries with a high commitment to research and development from relatively early on sought to secure a favourable position for themselves in the field of biotechnology. Whereas Germany as early as 1972 had started to develop an integrated national biotechnology programme holding open the possibility of direct intervention, followed by the Netherlands some ten years later, others lag far behind. Internationally, there are many technologically-poor countries from the Third and Fourth Worlds whose biodiversity or genetically-rich resources are attracting the kind of attention that threatens to subject them to the latest form of neocolonialism: what may be called bio-colonialism (see McNally and Wheale, this volume). Science continues to play a key role. The original visionary idea of determining the full chromosomal structure of humans became the central concern of an international scientific effort in the late 1980s. Supported by the US Congress, the European Union and the Japanese government, the Human Genome Project (Hilgartner, 1995) has set itself the goal of mapping and sequencing every gene by the early twenty-first century and thus completing a paradigm shift which will radically transform biology, biotechnology and medicine.

An adequate treatment of the politics of biotechnology would have to include an analysis of the relations between science, the universities, industry and the state (Street, 1992, pp. 86–7, 88, 125; Webster, 1990; Bijker et al., 1990), as well as relations between states (particularly the North and the South). Attention will also have to be given to conflicts arising from the employment of genetic tests for such diverse purposes as making health services more efficient through prevention, preventing and controlling occupational illnesses, streamlining recruitment to the labour market, reducing the disability burden on society through eugenic family planning, collecting and storing genetic data and thus altering the information infrastructure of society, etc. (van den Daele, 1986). Given that the concern of the present chapter is not the politics of biotechnology as such, but rather a discourse-theoretical clarification of the cultural form and social and political practices surrounding biotechnology, however, a wider perspective is adopted. This is what is nowadays often regarded as politics in the broad sense of contestation and conflict over cultural assumptions and social arrangements, or the very construction of society itself. Central in this case are the threats and dangers to nature and the human body more specifically, and hence to people's life-chances, but also to culture and social relations, perceived by the public as emanating from culturally defined social processes such as scientific research, technological development and industrial production (Beck, 1992). Threats and dangers of this kind are highlighted today by questions concerning the metamorphosis of collective goods into collective bads, the fundamental change in the relation of human beings to nature, the construction of nature and, even more radically, the possibility of the change, reconstruction and rational planning of human nature, the questioning of fundamental cultural premises, the unfathomable transformation of culture, not to mention the implications of all these for ecology, human dignity, the inviolability of the person, equality, human orientation systems and social relations.

The response of the public to biotechnological developments and to the concomitant push and pull exerted by science, industry and the state has undergone such transformation in the course of time that it is imperative to adopt as differentiated an interpretation as possible. Van den Daele (1992) is correct in highlighting the public's response to the cultural threat entailed by modern biology's turning of anthropological constants into variables, thus invalidating certain empirical assumptions about the *conditio humana*, and biotechnology's demonstration of our newfound capacity to completely reconstruct human nature. But one should be careful not to interpret public response, and the movements emanating from it, too strongly in terms of evaluative and prescriptive fundamentalism. Rather than focusing on particular moral and ethical orientations, one should note the marked differences at the institutional-cultural and political levels. It is indeed the case that genetics, genomics, genetic engineering and reproductive medicine mobilise religious feelings and provoke fundamental moral and ethical arguments and even legislation that are directed against biotechnology. Yet public response and social movements cannot without a trace be reduced to such fundamentalism. Far from simply languishing in an ethicisation or aestheticisation and moralisation of internal and external nature, the counter-currents are sufficiently differentiated to exhibit a political-ecological or political-technological orientation. Be that as it may, it is undeniably the case that the movements have played a central role in raising the question of nature, making risk into a problem, and placing the issue of the survival of the human species and its social form of life within the natural environment on the agenda. In so doing, they have made an indispensable contribution to the generation of the characteristic late twentieth-century discourse, the responsibility discourse, and hence the continuation of the discourse of modernity.

THE DISCOURSE OF MODERNITY

What both Habermas and Foucault have in mind when they refer to 'the discourse of modernity' is not the discourse that is of interest here, but rather an aspect of it. Both are concerned with what in Habermas's (1987) more precise terminology is called 'the philosophical discourse of modernity'. Their object of attention, which is admittedly analysed in their own peculiar ways, is a particular discourse with its own semantics, namely a philosophical discourse, that has been and is still being generated by the more encompassing discourse of modernity. The beginning of the discourse of modernity that is of interest here does not lie in the late eighteenth or early nineteenth century. Neither Kant nor Hegel, as Foucault and Habermas respectively maintain, inaugurated the discourse of modernity. It can be traced to the sixteenth century, particularly to the crisis of the Renaissance – to which can be compared the environmental crisis of our own time. During this period, a discourse ensued in response to the failure of the hitherto taken-for-granted understanding of reality to provide a shared background against which people could orient themselves and justify their activities. In contradistinction to a philosophical discourse, this was a societal discourse which provided the context for the development of a broad socio-political semantics in relation to the significant historical events of the time as well as a more specialised philosophical semantics.[2] The meandering course of this discourse, the discourse of modernity, can be followed from the sixteenth century to the present day.

'What Should We Do?'

The assumptions on which the sociological idea of the discourse of modernity is based admit of brief statement. Modern society is spanned by a permanently live network of public communication in the medium of which
collectively shared interpretations, definitions, meanings, knowledge and
even rational disagreements are developed and revised throughout the
process of its construction. This network becomes particularly activated
when, in the course of this ambivalent and contradictory process, the
shared background of taken-for-granted assumptions is interrupted, so that
order gives way to chaos, continuity to change, or certainty to uncertainty,
and the need arises collectively to identify, define and resolve a particular
societal problem. The discourse of modernity refers to this feature, so
characteristic of modern society, of agitated public communication around
a societal problem in which the collective activities of identification and
definition are discursively co-ordinated with a view to resolving them. It
thus links up with the experience and processing of social change and
transformation and the ensuing problems, but, in contradistinction to the
many local discourses continually under way, it does so at the macro-level
of fragile public communication where society forms an understanding of
itself as a whole, organises itself and takes action to determine itself. The
discourse of modernity is a societal discourse, the discourse of modern
society, that is generated at the macro-level by public communication and,
in turn, co-ordinates and organises that very communication.

Generally speaking, the problem at issue in the discourse of modernity
is the complex and only temporarily resolvable one of the constitution and
organisation of society. This raises the question of the ideality of the
social, and of the mutual compatibility, co-ordination, reconciliation and
consolidation of the different dimensions of society. What idea could
guide the constitution of society? How could identities, legitimately
defensible interests and differences most appropriately be organised under
prevailing conditions? What is the most appropriate common principle of
social identification for achieving this? How could deep-seated conflicts
be transformed into mutual understanding, rational disagreement or even
agreement? What collective political action is necessary to realise it? How
could the new arrangement be justified or rendered legitimate? The central
problem of modern society could thus be expressed in the form of
the pragmatic question, 'What should we do?' The resolution of this
deceptively simple question requires complicated social processes of co-
operation, conflict resolution, collective opinion and will formation, co-
ordination – in short, permanent discourse (Habermas 1992, pp. 56, 196).

This problem made its first appearance as a distinctly societal problem in the sixteenth century and, in its general thrust, has since become a defining characteristic of modern society.

The Master Frame

The manner in which this general problem addressed by the discourse of modernity has become collectively defined as an in principle resolvable one can be gathered from the cognitive structures or, rather, the cognitive order, that underpin the modern perception and experience of the social and natural world and direct and guide action,[3] and thus in turn again lend structure to the discourse of modernity itself. Since the beginning of the modern period, this cognitive order has taken the form of the macro- or master frame of free, equal and discursively structured social relations. It has allowed social actors from different forms of life and exhibiting different institutional-cultural characteristics to approach the problem of the construction of modern society from their own peculiar angles and to interact with one another and to act upon each other (e.g. through competition or conflict) in the pursuit of their points of view and interests.

It is only recently that the discourse of modernity has taken such a course that the liberal-egalitarian-discursive master frame of modernity has been modified and supplemented by additional principles or rules. Instead of organising perception, experience, action and communication in terms of social relations or society alone, the discourse of modernity was increasingly compelled through historical developments and events – e.g. the environmental crisis and the biotechnology revolution – and resulting experiences to incorporate phenomena that have a bearing on the relation between society and nature. Among these phenomena are the scientific objectification, the technological manipulation and the industrial exploitation not only of external nature but in particular also of internal (i.e. human) nature. The general problem addressed by the discourse of modernity thus became collectively regarded in terms of the society–nature master frame of the late twentieth century.

THE LATE TWENTIETH-CENTURY RESPONSIBILITY DISCOURSE

The problem at issue in the discourse of modernity first made its appearance in a societally significant form in the late sixteenth century in the wake of the Reformation, the Counter-Reformation and the ensuing crisis of the time. Throughout Europe, people began to realise that society brings itself into being and organises itself, that society not only requires its members to take the necessary collective action but also mobilises them to do so. Since it is of a general nature, this problem of the constitution and organisation of society never presents itself to be addressed as such. As a problem that repeats itself periodically in modern society as a collectively relevant one, manifesting itself under the influence of circumstances in an ever different form, it can be collectively identified, defined and dealt with only in the context of distinct, historically-specific discourses. The discourse of modernity can therefore be said to be produced and reproduced by a series of historically-specific but changing discourses. Under varying historical conditions, each discourse addresses the general societal problem articulated in the discourse of modernity by focusing on its specific manifestation and the collective political action necessary for its solution.

By considering the historically-specific cognitive structures or historical macro-frames constructed in the course of the production and reproduction of the discourse of modernity, one can grasp the particular problems addressed in the context of the discourses marking the different phases in the construction of modern society. Three historical master frames forming part of the modern cognitive order can be distinguished. They are the rights frame, the justice frame and the responsibility frame. In turn, they serve to provide a structure for the specific discourses through which the discourse of modernity is produced and reproduced. They are the sixteenth-, seventeenth- and eighteenth-century rights discourse, followed by the late eighteenth-, nineteenth- and twentieth-century justice discourse, and finally the late twentieth-century responsibility discourse.

In the following paragraphs, a brief characterisation of the late twentieth-century discursive context of biotechnology is developed by contrasting and comparing it with the two major preceding historical discourses.

The Rights Discourse

The first historical discourse, the early modern rights discourse, broke out in the sixteenth century in relation to the Reformation, the Counter-Reformation and the ensuing Wars of Religion, continued parallel to the Dutch struggle against the Spanish Crown and the English Revolution, and essentially ended with the completion of the American War of Independence and the first phase of the French Revolution in the late eighteenth century. The central institutional factor in this context was the absolutist state or, more broadly, the *ancien régime*. The rights discourse was carried by a series of debates, animated by the political or constitutional question, which were concentrated on the widely experienced problem of violence and disorder of the time and were unfolded in terms of the concepts of domination, sovereignty, resistance against tyrannical government, and formally recognised rights. In the course of this discourse, a socio-politically and culturally significant semantics was developed which was given coherence and consistency by a moral theory of rights. This theory, which gave concise formulation to the frame structuring the discourse, provides the justification for giving it the name of the rights discourse. In addition to forwarding a solution to the collectively defined societal problem, the rights discourse also contributed to identity formation and the mobilisation of the collective political action necessary to realise the proposed solution. While mobilising other actors such as the state, it particularly assumed the form of the early modern social movement – what is commonly called the classical emancipation movements. Having negotiated its course via a protracted series of exchanges and conflicts, the rights discourse eventuated in the establishment of a new, law-based institutional infrastructure that was assigned the task of taming violence and creating order in society. It took the form of the constitutional state.

The Justice Discourse

The second of the three historical discourses of modernity arose and developed in response to the market-based industrial-capitalist economic system and the effects it generated. It first appeared in late eighteenth-century England, rapidly spread throughout continental Europe in the nineteenth century, and was reproduced in a sublimated form under the conditions of the welfare state into the second half of the twentieth century, when this institutional arrangement started to break down. The justice discourse was activated by a series of debates which gave expression to the social question and are often collectively referred to as the poverty debate. The reason for this is that the problem complex that provided the focal point of this discourse was the exploitation, pauperisation and loss of identity that followed in the wake of industrial capitalism. Initially, all the participants accepted that the objective laws of the new system were unalterable, yet in a drastic turnabout in the discourse during the latter part of the nineteenth century the problem was collectively defined as being resolvable by means of state intervention on the basis of scientific-technical progress. This solution took the form of state-supported industrial-capitalist production and welfare through the

Table 1.1: Discourse of modernity: Biotechnology and the responsibility discourse in context

Period	16th–18th century	late 18th–mid-20th centuries	late 20th century
Historical events	Dutch Revolt, English, American and French Revolutions	Industrial Revolution, World War I & II	Biotechnology revolution, Environmental crisis European revolution of 1989
Society	Early modern	Industrial capitalist	Risk or cultural?
Discourse	Rights	Justice	Responsibility
Problem	Violence	Poverty	Risk
Question	Political or constitutional	Social	Nature
Issue	Survival of society in its political environment	Survival of society in its social environment	Survival of society in its natural environment
Master frame	Rights	Justice	Responsibility
Identity	Liberalism	Socialism	Environmentalism
Collective actors	Monarchy; aristocracy; bourgeoisie; classical emancipation movements	Capitalists; state functionaries; labour movement	Industry; state functionaries; new social movements
Institutional infrastructure	Constitutional state	Welfare state	Neo-corporatism or post-corporatism(?)
Means	Law	Money	Knowledge

redistribution of the fruits of growth. On the one hand, this discourse gave rise to a socio-politically and culturally significant semantics that was lent consistency by the logical rules of a new moral theory, the theory of justice. On the other, it contributed to appropriate identity formation and collective political action by making possible the construction of the working class or labour movement as a collective actor. The impact of the justice discourse can be gleaned from the significant degree to which the existing institutional arrangement, the constitutional state, was modified and supplemented by a new departure, namely the money-based institutional infrastructure known as the welfare state.

The Responsibility Discourse

I propose to call the current discourse producing and reproducing the discourse of modernity the responsibility discourse on the basis of the observation that the historical master frame of the preceding period – the justice frame – is being relativised by another – the responsibility frame – and that, consequently, the theory of justice is being recontextualised by a new semantics.[4] The latter takes the form of the moral theory of responsibility.[5] This moral theoretic or philosophical semantics gives expression to the new master frame and articulates the internal coherence and consistency of the currently emerging historical discourse. Since the 1970s, the responsibility discourse has been generated by a variety of intertwined debates, which all, in one way or another, deal with what has become known as 'the problem of nature',[6] which means to say with some aspect of the relation between society and nature, both external and internal. They range over a series of apparently disparate problems, including science and technology, particularly the life sciences and biotechnology, industry, particularly its nuclear, chemical, pharmaceutical and genetic sectors, collective decision-making, the status of collective goods, gender, the environment, etc. Given that it has played a part in the discourse from the outset and that it has progressively assumed increasing significance, the concept of risk (Beck, 1992; Evers and Nowotny, 1987) has gained currency as a means of making sense of these problems. This concept captures the threats, dangers and disadvantages that culturally defined social processes harbour for collective forms of life. At the same time, it also invokes its complement: the responsibility that needs to be taken for the collective definition and social organisation of such processes as well as their products and especially their consequences. As far as collective political action aimed at redressing these problems is concerned, the responsibility discourse has made possible the deconstruction of older forms of collective action such as the labour movement and the construction of novel identities and forms of collective action collectively known as the 'new social movements'. Of particular interest in the present context are the new anti-biotechnology movements (see e.g. McNally and Wheale, this volume). As a result of the exchanges, conflicts and negotiations forming part of the responsibility discourse, a new society with a knowledge-based institutional infrastructure is at present coming into being which would seem to possess some novel features. No agreement has yet been reached on how this society and its new institutional order should be characterised: the 'risk society' (Beck, 1992), 'cultural society' (Moscovici, 1968; Lash, 1994), 'neo-corporatism' (Willke, 1992), or 'post-corporatism' (Eder, 1996a)?

The Responsibility Frame

A public discourse is generated through the communication by different collective actors of competing and even conflicting definitions or framings of a common problem. The outstanding contemporary example of risk communication borne by social movements, science, the state and industry is comparable to historically significant examples such as violence communication and poverty communication. It is through such conflictual yet structured communication that a common cognitive structure in the form of a macro- or master frame gradually becomes established, which in turn restructures the communication and hence the discourse itself. A whole new cognitive order or system of classification of reality takes effect. This is the case today with the responsibility frame, as it had been with the rights and the justice frames in previous eras. In relation to such a cultural development, the actors involved acquire new ways of thinking, feeling and evaluation, new concepts, behavioural dispositions and norms. The master frame thus makes possible but also draws the limits of identity formation, collective mobilisation and action and eventually the ensuing institutional innovation and change.

In a manner not unlike the rights frame and the justice frame, which respectively emerged and became established in the late sixteenth and early seventeenth centuries and in the late nineteenth century at a critical turning point towards which the preceding phase of breakdown and loss of certainty had been heading relentlessly, the responsibility frame made its appearance in the 1970s. Comparable to the 1570s and again the 1870s, these years mark the beginning of a turning point, a discontinuity to which biotechnology is central, in the wake of which a new master frame or cognitive order emerges and starts to become established. A discontinuous period of this nature typically indicates a transition between the two major phases of a public discourse. Taking cues from the rights and justice discourses, the first phase is typically characterised by the breakdown of hitherto taken-for-granted assumptions and consequently the prevalence of a high level of uncertainty and unregulated or at best diffusely directed conflict between irreconcilable points of view. In the second, by contrast, foundational assumptions are reconstituted and a new certainty is slowly re-established, with the result that disagreement and conflict are channelled in a more constructive direction. Between the two phases a change of the greatest moment typically takes place in that the cognitive order is more or less fundamentally reconstituted by the establishment of a new, macro-cognitive structure or frame. The responsibility frame is precisely such a master frame which provides a new set of assumptions, a new mode of classifying reality and a new range of interpretations of the situation. At the same time, it also makes possible the projection of a collectively accepted vision of a solution to the societal problem of risk, including the appropriate regulation of biotechnology. As such the responsibility frame is a major factor in enabling contemporary society to weather and, if it is not already too late, to pass through the debilitating environmental crisis that has been afflicting it for some time. Through the new confidence and certainty arising upon its basis, the responsibility frame is enabling the late moderns not only to form a new identity, or rather, a range of different yet closely related identities, but also to mobilise collectively for social change and to engage in collective action of various kinds to bring about the envisaged change. The responsibility frame is a common, or at least a widely shared, new macro-cognitive order, which effects a shift from uncertainty to a new certainty, yet without obliterating a plurality of competing and even conflicting meso-frames. As in the past, when absolutists, republicans and constitutionalists clashed over rights, or when capitalists, socialists and liberals conflicted over justice, so today biotechno-science, industry, the state and the new social movements, including conservationists, fundamentalists and political ecologists and technologists, are struggling over responsibility. The resultant discursive organisation of this competition, conflict and struggle is centrally dependent on the responsibility frame.

The most important question that remains, and which I wish to answer at least in a preliminary way here, is what an analysis of the responsibility frame would reveal.[7]

To begin with, it is, like any frame, a cognitive structure or set of rules which allows human beings to classify the various phenomena in their world, thus organising their experience and interpretations of nature, society and themselves. This it does by enabling them to give meaning to their world, to think about or conceptualise it, and finally to regulate their behaviour in relation to it. In the case of the early modern rights frame, this basic classification turns on the constitution of society by means of politics, whereas by contrast in the later justice frame it turns on society as such and in the contemporary responsibility frame on society as a part of nature. As such, the responsibility frame consists of the following structural elements or rules: (i) human beings are born free and are endowed with reason and hence they are becoming what they are: free and rational beings; (ii) not only human beings but also civilisations are essentially the same, the human universal being language and communication or codes; (iii) humankind as an idea has become a fact, just as history has become world history, but instead of the particular being subsumed under the universal, humankind is exemplified by everyone's here and now, by the local or place; (iv) for the first time in history human beings have to assume planetary responsibility; (v) it calls for realisation through a macro-ethical commitment of collective coresponsibility, thus leading to the creation and the maintenance of a communication community; (vi) the objective framework for this self-constitutive, self-organising and self-maintaining communication community is nature in the sense of both the body and the planet as its home; and finally (vii) practical reason is primary for the self-constitution and self-organisation of embodied, planetary society.

It is these elements, structures or rules making up the responsibility frame that structure the contemporary discourse, lending it its internal coherence and consistency, and thus make possible the discursive organisation of the communicative conflict surrounding biotechnological risk. While they find their most explicit and systematic formulation in the writings of moral philosophers such as Hans Jonas and Karl-Otto Apel, these rules – and hence the responsibility frame – are a special quality that pertains to the communicative and discursive practices generating the responsibility discourse. They represent what is called reflexivity. The responsibility frame kindles, keeps alive and even imposes an awareness of the cognitive, normative and conative dimensions of the late twentieth-century cognitive order, and it constantly directs attention towards the effects of classifying and dealing with reality, including ourselves, in those terms.

The civilisation of the gene is fraught with dangers, threats and risks, but simultaneously it also harbours a promise. During the past decade or two we have been witnessing the birth pangs of a new historical consciousness and mode of existence. Instead of a universalism realised by a dynamic philosophy of history, we are increasingly compelled to become reflexive about and to confront generality – from the genetic codes of our bodies through human language and communication to humankind and its fragile conditions of existence.

2 Biopolitics in the Risk Society: the Possibility of a Global Ethic of Societal Responsibility

Gerard Delanty

In this chapter I explore two related themes: the significance of new forms of technology, in particular biotechnology, as a political-ethical challenge in what Beck (1992) has termed the 'risk society'; and the ability of contemporary 'biopolitics', to use a phrase of Feher and Heller (1994), to generate an appropriate ethical response. The argument I shall be proposing is that an adequate ethical and political response to new forms of technology in the 'risk society' must abandon essentialism as well as questioning a simple model of rights as a normative reference point. The crucial question, which will be addressed in the third section, is how such an adequate response must draw upon an ethic of societal responsibility (see Strydom, this volume). This is approached by means of a debate with the theories of Jonas, Apel and Habermas. These theories allow us to see how societal responsibility provides a new normative framework in which debates and struggles on biotechnology take place. But the new discourse of biopolitics has wider implications for society: it challenges our ideas of democracy and the ethical neutrality of science. As a key factor in motivating contemporary perceptions of risk, biotechnology opens a crucial discursive space in which critical communities are and can be mobilised to specify problems and to propose solutions. My argument, then, is that as a result of new developments in biotechnology, a new kind of politics is advanced: biopolitics, which involves a reconstruction of the politics of science. In this process, science's monopoly over the definition of what should count as scientific rationality is broken and many voices emerge around the theme of societal responsibility. Accommodating these differences would appear to require a more inclusive and participatory democracy, one that is centrally concerned with the relationship between society and nature, thus supplanting liberal-welfare democracy, which is about the relationship between society and the state through delegated represesentative models.

TECHNOLOGY IN THE RISK SOCIETY

To the extent to which technology figured in the philosophies of modernity it was generally seen as a force of potential liberation from nature. Technology was the means by which society humanised nature. Modernity implied the technical mastery of nature. This view of nature and the civilising power of modern technology was the expression of the nineteenth century's great faith in science and human progress. That technology could undermine the ethical and political fabric of society was unimaginable. As a means to an end, technology was simply part of the great narratives of progress and enlightenment. Those who rejected technology also rejected modernity for an imaginary premodern age of harmony. Aside from romantic reactionary philosophers, the great proponents of modernity – radicals, liberals, socialists – did not question the role of technology in the march of progress.

This, of course, is not entirely surprising because the modern condition was born in the age of the Enlightenment at a time when the enemy of society was the state. For the great proponents of civil society and modern rationalist politics, the absolutist state was the enemy of a free society. The new age of emancipatory politics promised a civil order based on rights and a social contract by which the state was bound to acknowledge constitutions and declarations of human rights. Civil and human rights were seen as the means to protect society from the state. Thus was born the modern idea of negative liberty, the freedom *from* the state. The social contract, the bond between the individual as a member of civil society and the state, perfectly expressed the modern condition and its two great discourses of rights and justice. The ideas and ideals of the Enlightenment assumed that sovereignty resided in civil society and not in the elite world of the court society. Moreover, when this notion of sovereignty was extended to refer to a territorial realm, the idea developed that the nation-state was the protector and the political form of civil society.

Until the advent of the nuclear bomb and the spectre of world-wide destruction this simple vision of sovereignty and the social contract was taken more or less for granted. It was assumed that once the institutions of civil society held sway over the state all was well with the world. The spectre of nuclear war challenged this, though many still sought comfort in the fact that military weapons were firmly under the control of the responsible state, which in turn was answerable to civil society. Only in recent years has a new spectre emerged: the spectre of the risk society. With the decline in the threat of a nuclear war between the superpowers, the risk society has moved to the fore of the political imagination of the post-Cold War era.

Modern society was based on the assumption that technology was a means to an end and that the end was human progress and emancipation. This is the 'tool' model of technology: technology as a thing. But with the growth of new forms of technology, it has increasingly become apparent that the means have not only outgrown the ends, but that they have become potential sources of destruction. New forms of technology differ from older forms in that they cannot be readily identified as visible points on a linear process which can be simply subject to regulation: they have become diffuse and recalcitrant. Technology is no longer a mere thing, but has become part of the social fabric itself. Moreover, the view can no longer be seriously entertained that technology is neutral and that evil resides in society. Technology has now reached a point where it is possible for it to construct nature and remould the environment. These developments raise two major issues, both of which have ethical implications: first, the disconnection of technology from normative systems of regulation; and second, the growth of societal complexity.

It is increasingly becoming apparent that the logic of scientific diffusion is one that is not dependent on normative control and that we are reaching the age of 'anti-paradigms'. What this means is that the Kuhnian notion of a scientific paradigm, in which normal science is conducted under the aegis of a community of consensus, may no longer apply and that science may be becoming the enemy of society. The enemies of the 'open society', which Popper defended, may not in fact be ideologies and forms of state power but science itself: the new spectre of 'piecemeal engineering' rather than utopian grand designs has now become the reality with which society is faced. Beck (1995a, p. 9) points out that science is now internally divided and is continually contradicting its own safety claims, and consequently the authority of science on safety matters has been eroded.

Another problem with which society is confronted is that of complexity. According to Luhmann (1981a) modern society has now reached a level of differentiation to the extent that it is no longer possible to speak of society having a 'centre'. Complexity is one of the most apparent characteristics of late modern society, and according to Zolo (1992), we cannot assume a simple, transparent situation in which there is a single power centre capable of reversing adverse developments. The social and ethical implications of these developments is that the new class of technocrats do not in fact exert control over the technological apparatus to the extent to which it is comforting to imagine. Complexity is not only a feature of society in general, but has also penetrated the constitutive sub-systems of society themselves. Science and technology have reached a degree of complexity where not even the most highly skilled specialists in particular areas have more than basic knowledge of other areas.

According to Ulrich Beck (1992), late modernity has institutionalised 'risks': the social contract, which was the basis of high modernity and which guaranteed that society was safe from danger, has been invalidated by the recalcitrance of nature. The social contract and its discourses of rights and social justice – from the struggle to create the first constitutions to the debates on the distribution of the social product under the welfare state – are now being rapidly overtaken by the recognition of a new discourse. A concern with social needs – to be alleviated by economic growth – has shifted to a concern with risk. In the face of the threats posed by new forms of technology we are becoming aware of the inadequacy of rights and justice as frames for the protection of society. The ideas and ideals of modernity promised the present age a secure future; but this is now being threatened by technology itself. Modernity was confronted with dangers – the dangers of an uncontrolled nature; late modernity is confronted with the risks produced by technology. Previously these risks were local and products of external agency; today they are universal and creations of human agency. These risks ironically are products of modernity's struggle with danger. Weber's concept of rationalisation is unable to grasp this late modern reality: 'Along with the growing capacity of technical options grows the incalculability of their consequences' (Beck, 1992, p. 22). The older critiques of the consequences of instrumental reason and the domination of nature, such as the argument of the *Dialectic of Enlightenment* (Adorno and Horkheimer, 1979), claimed that the consequences of technology lay in human subjugation. The problems of the risk society in fact lie not in the instrumentalist striving towards an end by means of maximising the means, as Adorno and Horkheimer assumed, but in the dangers associated with the means themselves. The consequences and side-effects are now the problem. The normative project of the risk society, Beck (1992, p. 49) argues, is safety: 'The commonality of anxiety takes the place of the commonality of need.' The welfare-democratic state has become a 'safety state'.

The risk society has radicalised the principle of falsification, the 'trial and error hypothesis', which Popper (1974) believed characterised the logic of science. In the risk society (Beck, 1995b, p. 106) 'all accidents and disruptions – for instance, in nuclear power plants all over the world –

are experimental findings in a continuing, perhaps undecidable concrete experiment.' What Popper did not anticipate is that society has now become a gigantic laboratory for science to conduct its trials and errors. The method of 'critical rationalism' – that science must be methodologically self-critical because it is constrained by the fact that the pursuit of scientific knowledge can never be complete as it is forever open to falsification – does not compel science to enter into a critical relationship with the object of research. It merely asserts that scientific knowledge and progress come from the process of learning from errors. But the reality is that these necessary 'errors', which in Popper's view constitute the core of science, and their accumulation, which adorns the progress of science, have today become social 'errors'; in short, they have become risks.

Biotechnology is a particularly interesting example of the kind of technology and scientific rationality that has come to prevail in the risk society. Biotechnology has undermined one of the premises of the Enlightenment: the idea of objective nature and that technology and science can render it socially useful. Biotechnology does not simply mediate between nature and society; it shapes nature in its own image and in doing so poses moral dilemmas for society as a result of the hazards it engenders. The greatest myth of all about technology has been exploded by biotechnology: the myth of the neutrality of science and the idea that 'ought' cannot be derived from 'is'. One of the most pervasive myths about science and technology is that they are neutral and society is responsible. Only society knows what is right and what is wrong: science is value-free. Recent developments in biotechnology have demonstrated that even gender, once thought to be immutable and part of the fixed order of nature, can be shaped in the image of modern science. Since the discovery of the DNA structure in 1953 and the development of recombinant DNA engineering, such as the ability to correct genetic errors that cause hereditary diseases, the patenting of life and the ability to introduce alien DNA into the cells and chromosomes of animals, it is no longer possible to assume that nature lies outside human agency. Nature itself has become a construction of technology. The release of genetically modified organisms into the environment, the cloning of livestock according to supermarket specifications and the prospect that it is now possible to clone human beings undermine the coherence of the very notion of nature. For Beck (1992, pp. 80–1) we have reached 'the end of the antithesis between nature and society. That means that nature can no longer be understood outside of society, or society outside of nature ... At the end of the twentieth century nature is society and society is also "nature".'

Hannigan (1995, p. 175) has pointed out that concern over bio-technology does not originate *within* science but rather is a critique of science. It would appear that science and technology are no longer the ideologies that Habermas (1971) attributed to them. Science has lost a considerable degree of its own self-legitimation as a result of the rise of anti-technological currents in society as well as the scientisation of protest and new demands that science be made accountable to the media, government and the public. As Beck (1992, p. 29) argues, 'in definitions of risk ... science's monopoly on rationality is broken' – for in questions of risk there can be no expert. What emerges instead is a new area of contentious action. In the continued experimentation with high-risk technology new social actors encroach on the field of scientific rationality, merging it with communicative reason – the normative question of how we wish to live. Thus social actors claim the mantle of science while others reject science in the name of essentialist models of truth. Beck (1995b, p. 106) claims that in this situation the 'boundaries between production, research and use are blurring', and, as a result of the diffusion of contentious action into the core of scientific experiment,

> many voices are involved in this experiment. Differing, antagonistic worldviews appear. The technical people are interested in technical success, which may well be independent of health effects, social and political turbulence and responsibilities, and lest we forget, the verdict of profitability. A chorus of voices and viewpoints argues about the course and outcomes of such experiments.

An issue I now consider is how new social movements have responded to the forms of technology that have accompanied the risk society. The question is whether the risk society and its products, such as biotechnology, challenge the oppositional culture of new social movements even as these movements challenge the acceptability of its potential risk.

BIOPOLITICS IN THE AGE OF RISK

In the previous section I argued how, in contemporary perceptions of risk, debates on biotechnology come to the fore in breaking down the Enlightenment framework of the ethical neutrality of science. In the risk society the dualism of science and nature collapses. Consequently, science can no longer be perceived to be a discourse lying outside society. In particular with respect to biotechnology, the implications of this levelling of science for politics are far reaching: in the risk society politics loses its focus on the 'Other'. New reference points move into the foreground: democratic accountability and societal responsibility. As Beck (1992, pp. 36–8) has argued, in the risk society risks are democratic – all classes and social groups are equally threatened – so the response to risk must also be democratic and reflect the new kinds of power, the 'organised irresponsibility' of science. Thus biopolitics will be largely articulated around the politics of knowledge; the politics of the definition and legitimation of risk.

Ferenc Feher and Agnes Heller in *Biopolitics* (1994) have provided new terms of reference for the debate on political ethics. By means of the concept of 'biopolitics' they argue that nature, which had been banished by the Enlightenment from the political, has returned to the centre of the political and today constitutes a new biopolitical subject. Biopolitics is a politics that lies beyond the state and expresses the end of the quest to construct macro-political agents. For the most part it does not have an 'institutional imagination' (p. 34), and has pitted itself against mainstream politics, even if it occasionally makes alliances with legislative bodies. The grand narratives of class-based politics, nationalism and the sexual liberation movements of the postwar period have run their course and with the *fin-de-siècle* spectre of end of the Cold War – AIDS, a new racist malaise, civil wars – the political has sought to express itself in new language-games: the politicisation of the body as articulated through the language of difference. The new biopolitics of health, race, gender and environment, in the view of Feher and Heller, is part of the reaction to the 1960s, and a major part of it is counter-revolutionary (p. 62). Ironically, while mainstream politics has begun to abandon the 'grand narratives', various contingents of biopolitics are now about to elaborate their own versions of the same (p. 39).

Biopolitics possesses the potential to transcend the traditional 'friend'/ 'foe' polarity, but is double-edged. On the one side there are the militants, the speech-police, for whom only the majority is racist while they themselves can be free to express hatred against the Other. In this instance biopolitics becomes a politics of 'self-closure', an 'ontological fall-back' (p. 27). On the other side, there is the possibility that the new wave of biopolitics can transcend the old ambitions of radicalness which was always to destroy utterly its opposite (pp. 32–3). What is at stake is a fundamental choice between a biopolitics of 'self-closure' or one of 'dialogical politics' (pp. 24–5).

The problem Heller and Feher raise is whether the 'biopolitics' of the new social movements is really capable of providing solutions to the challenges of contemporary society. Two models of biopolitics emerge from their analysis. Biopolitics as oppositional, an adversarial politics which appeals to notions of rights and cultural essentialism, and a dialogical one that points to a new democratic universality. In the present context, only a model of biopolitics that is capable of articulating a non-adversarial kind of politics will be able to meet the challenge of biotechnology. This latter model is also reflected in Beck's (1992, pp. 37–8) argument that the risk society offers a basis for a new politics for the simple reason that risks are 'democratic', affecting both rich and poor. By means of the 'boomerang effect' the risks catch up with those who produce them, creating a situation in which perpetrator and victim sooner or later become identical. Beck (1995b, pp. 33–4) believes the politics of risk could become the basis of a new ecological enlightenment which could move into the political space created by the end of the Cold War. In this vein Eder (1996) argues that ecology in its post-environmental phase has ceased to be an oppositional movement and has become a new master frame in public discourse which is addressed by all social actors and has even supplanted conservatism. Biotechnology is an example of this in that it is a discourse that is not dominated by any one social actor who has a monopoly over ideology. It is also an example of the obsolescence of oppositional ideologies for it is not something reducible to state power or hegemonical domination. Consequently oppositional politics alone is helpless.

The foregoing discussion has questioned the adequacy of both cultural essentialism and straightforward declarations of rights as a basis for a biopolitics in the age of risk. New forms of technology such as biotechnology cannot be simply counteracted by appeals to rights, be they those of human beings or of nature. The retreat of ecologically oriented politics into essentialistic stances results in a moralising of politics. Politics is not reducible to morality; it transcends the 'self-closure' that is inherent in much of the identity politics of biopolitics. The problem is not so much the question of collective identity, as the reducibility of politics to morality.

The argument defended here is that neither a liberal view of rights nor a culturally constructed view of nature as an essence can offer an adequate challenge to such new forms of technology as, for instance, biotechnology. The liberal model of rights assumes that fundamental rights are self-evident, prior to society and can be legally institutionalised. This model assumes, furthermore, that political contention is merely about the empirical application of those rights, which are supposed to be clearly defined. A rights model takes for granted that the 'common good' is both common and good, and that it can be maximised by the legal institutionalisation of rights. The problem with a rights model is that rights in the area of environmental discourse are simply subject to so much contestation that it is not possible for them simply to be subject to legal institutionalisation. Environmental debate is frequently conducted on the level of conflicts over basic values. In feminism too, it has been argued that a rights discourse has its limits (Kingdom, 1991).

On the other side, the fundamentalist trend in the biopolitics of new social movements assumes too great a coincidence of politics with morality: biopolitics reduces collective action too much to identity politics. Biopolitics has a strong constructivist element to it: ecological protest is as much about reconstructing the social as about the risks themselves. Thus powerful symbolic rhetoric (the purity of nature as embodied in milk) acquires a strong adversarial quality (Hannigan, 1995, p. 176). Because the opposition is against science, opposition does not always rely on scientific foundations for its arguments. The result is that, while collective identities may be shaped, no effective action is undertaken. In the social construction of risk in the new social movements, the objectivity of the actual scientific status of the risks is marginalised in what is essentially a politics of identity. This sense of social constructivism is apparent in Beck's analysis of how cultural symbols and social experiences govern the construction of risk. He (1995a, p. 55) argues: 'The ecological movement is not an environmental movement but a social, inward movement which utilises "nature" as a parameter for certain questions.'

Beck (1992, p. 55) goes so far as to admit: 'It is not clear whether it is the risks that have intensified, or our view of them.' This is the basic contradiction of Beck's idealist version of constructivism (Barnes, 1995, pp. 106–11). Constructivism is caught in the bind that to admit that the risks are really there would be to admit the validity of scientific knowledge; but the problem is that it is scientific knowledge itself that is being challenged by the new politics (Murphy, 1995).

To pose the problem in slightly different terms: how are we to conceive of the relation of the ethical to the political? If rights – along with the other aspirations of industrial society such as the social question – do not provide the answer, and if biopolitics is also unable to come up with a means of articulating a politically applicable ethical framework appropriate to the challenges of the risk society, can a discourse of responsibility offer an alternative? This will depend on freeing the notion of responsibility from some of its traditional limitations.

Traditional ethical concepts of responsibility were based on the idea of individual responsibility. On the other side, the dominant understanding of politics has been the macro-subject, the social movement, for instance. A genuinely political ethics, then, does not exist as such. The issues Feher and Heller raise provide important terms of debate for developing a theory of political ethics. Biopolitics as a manifestation of a new kind of political ethics bridges the gap between ethics – the sphere of individual responsibility – and politics. Rather than retreat into the fundamentalist politics of cultural essentialism, the new biopolitical space should focus on a political-ethical universalism. Such a universalism, if it is to be non-foundationalist and anti-essentialistic, can only be discursively mediated.

In the present context – social responses and policy implications of biotechnological innovation – what this discussion opens up is that nature is itself a social and cultural artefact. There is no natural nature (Soper, 1995). Any attempt to change it is a political venture and therefore calls for a process of collective mediation, a dialogical politics as opposed to a politics of self-closure. In order to clarify what a discursively mediated notion of societal responsibility consists, I turn now at the theories of Jonas, Apel and Habermas. Jonas was the first major philosopher to discuss the concept of a post-Kantian ethic of responsibility, and while Apel attempted to medate the idea of societal responsibility with a communicative theory of ethics. Habermas has extended this discussion with his discursive theory of democracy.

THE DISCURSIVE FOUNDATIONS OF AN ETHIC OF SOCIETAL RESPONSIBILITY

The problem with traditional ethics is the assumption of proximity and visibility. It was always assumed that the individual was the subject of ethical action. In practical terms this amounted to the assumption that the secondary consequences of actions could be both identified and controlled by the moral agent. The classical model of this is Kantian moral theory, which stresses the autonomy of moral reason. In the present context of debates over biotechnology there are many problems with this. As Hans Jonas (1973, 1974, 1976, 1984, 1994) and Karl-Otto Apel (1988) have argued, the overriding need today is for a global ethic of responsibility capable of addressing technology. Jonas's argument is that a new notion of responsibility must extend beyond contemporaries to future generations. This is an insight that cannot be accommodated in the traditional model of responsibility, which referred to the here and now. For Jonas (1984, p. 8), a new ethic of global responsibility must address the very survival of life itself; it is a responsibility not just for human life but also for nature: 'No previous ethics had to consider the global condition of human life and the far-off future, even the existence, of the race. That these are now at issue demands, in brief, a new conception of duties and rights, for which previous ethics and metaphysics provide not even the principles, let alone a ready doctrine.'

The Kantian model was too simple. It assumed that we are responsible only if we have the capacity to act otherwise. This in turns presupposes proximity: we are responsible only to those in our immediate proximity and for action over which we have control: 'The ethical universe is composed of contemporaries, and its horizon to the future is confined by the foreseeable span of their lives' (Jonas, 1974, p. 7). For previous ethics, good and evil were close to the act and were not matters that require remote planning (Jonas, 1984, pp. 4–5). But the problem is that technology has now created new kinds of situations which make proximity meaningless. Technology, in particular biotechnology, has itself created situations that transcend proximity and can even manipulate nature, which in the Kantian system was outside the moral sphere. Biotechnology undermines one of the taken-for-granted assumptions of previous ethics: that nature consists of an unchanging essence. The consequences of technology do not emanate in an obvious way from the decisions of individuals; they are more complex than can be accommodated in an individual model and pertain to the logic of science itself and the social conditions of human actions. The knowledge that is required to make ethical decisions is no longer that available to the solitary individual, but encompass science itself. There is in short no 'categorical imperative'. Unless a new conception of morality can be found there is a danger that society will become the object of technology, for technology has long since ceased to be an object or a tool itself.

Jonas argues that the Kantian 'moral imperative' must be replaced by a new 'imperative of responsibility'. The 'new imperative' is addressed to public policy rather than private morality' and involves a very different kind of consistency from that of Kant: 'not that of the act with itself, but that of its eventual effects with the continuance of human agency in times to come' (1984, p. 12). What has changed, then, is that the spatial and temporal horizons of ethics have been radicalised: in place of contemporaneity and proximity, the global order and future generations has appeared. 'Modern technology has introduced actions of such novel scale, objects, and consequences that the framework of former ethics can no longer contain them' (Jonas, 1974, p. 8). In Jonas's view (1984, pp. 3–4) an ethic of responsibility will have to break from the tradition of the city, which was the sole domain of the previous model of responsibility. But it was a kind of responsibility that did not extend to nature and was limited in scope: 'It is in this intrahuman frame, then, that all traditional ethics dwells, and it matches the size of action delimited by this frame.'

Jonas's critique of technology is not new to the extent to which he sees modern technology as a supreme danger. This was a view that was deeply ingrained in much of German philosophical conservatism, of which Heidegger and Schmitt were the supreme examples, and is also reflected in the writings of Lewis Mumford. What is interesting about Jonas's work, which has now become a classic, is that he has proposed an ethics of global responsibility as the only viable ethical response to modernity. This new ethics is supposed to be capable of bridging the gap between politics and ethics. However, it is not entirely clear how this is actually to come about. There is the strong suggestion that Jonas is in favour of an alliance of knowledge and power, a position that comes close to endorsing the culture of experts instead of public participation in debates (Bernstein, 1994, p. 845). Collective responsibility cannot be reduced to being an 'imperative' as Jonas believes. At this point the question of institutional-isation and the reopening of the political around public discourse arises. Rather than take the discussion on responsibility in a postmodernist direct-ion (Smart, 1995; Tester, 1993, pp. 102–26; White, 1991, pp. 19–23), a focus on communication may offer an alternative path. Jonas himself is very vague about such questions of the political. Above all, what is missing from his analysis is a notion of communication.

Karl-Otto Apel (1978, 1980, 1984, 1988) offers an interesting perspec-tive on this. In his view the problem is to overcome what he calls the 'system of complementary of western ideology'. This is the separation of public from private morality. The norms that guide the public sphere (politics, law, economy) are regulated by the value-neutrality of science and technology. In this sphere ethics is merely a matter of simple major-itarianism and decision-making according to agreed procedures. On the other side, morality is largely a private matter. The problem that this pre-sents, according to Apel, is that there can be no moral responsibility for the collective consequences of social action. It is precisely this – the con-struction of an ethical-political 'macro-ethics' – that is the contemporary task. In order to achieve this task, the aim should be a radical questioning of our ethical and political dualism. Apel, however, has mostly confined his analysis to a highly normative level and does not address issues relat-ing to the social basis of macro-ethics. His notion of the 'communication community', however, is an important step in the direction of a discursive theory of democratic participation. Ethical disputes, Apel has persuasively argued, can be resolved only by communicative reason and not by refer-ence to an absolute point of justification. Human communication itself provides the normative structures for conflict resolution.

Undoubtedly one of the most advanced statements on the social possibility of a new political ethics is that of Habermas's discourse ethics (Habermas, 1993, 1994, 1996). It would be beyond the aim of this chapter to discuss this in depth, but I would like in conclusion to point theoretically to a way forward suggested by Habermas's notion of discourse. One of the most fruitful implications of Habermas's model of a discourse ethics is its extension into discursive democracy. Like Jonas, Habermas rejects the philosophy of consciousness as a normative model and proposes a discursive politics of conflict resolution and a self-organising legal community. Habermas's model defends the case of a new politics of public opinion and debate. Politics can no longer be left to expert cultures; and on the other side, morality cannot be confined to the private realm. How does this pertain to the political-ethical challenges of biotechnology?

The political implications of Habermas's discourse model have not been discussed to a great extent (Blaug, 1996; Dryzek, 1990; White, 1995). It is fruitful to consider how they might be applied to new forms of technology. In his most recent contribution to discourse theory, Habermas (1996) has recast his older theory of the public sphere (1989; Calhoun, 1992) into a theory of discursive democracy linking notions of legitimation, civil society and the public sphere together. The value of Habermas's approach is that his notion of discursive democracy offers an alternative to simplistic notions of participatory or direct democracy. Discursive democracy recognises that in a complex society the capacity of the public sphere to solve problems on its own is limited; it must leave many issues for specialised treatment (Habermas, 1996, pp. 359–60). The role of the public sphere, rooted in the associational network of civil society, is one of 'problematisation'. According to Habermas (1996, p. 373), in complex societies the public sphere consists of an intermediary structure between the political system and the private sectors of the life-world and functional systems. In his view (1996, p. 351), 'it is advisable that the enlarged knowledge base of a planning and supervising administration be shaped by deliberative politics, that is, shaped by the publicly organised contest of opinions between experts and counter experts and monitored by public opinion.'

What is to be questioned – and it is here that biotechnological communication touches on the core questions of political communication today – is the equation of discursive democracy with a theory of participatory democracy modelled on the fabled *polis*. Given the scale of decision-making and the complexity of modern society, radical democracy must be rescued from the illusion of the *polis* and its traditional model of responsibility. A new ethic of responsibility will have to be pluralistic and discursive; it will have to recognise that there is not one public sphere but many, and that different kinds of discourse may be operative in them.

A more differentiated model of discourse would be one that recognises that the simple participatory model is unrealistic and in its place there must be a plurality of public spheres: expert cultures interacting with intellectual and oppositional cultures. The democratisation of responsibility must extend to a wide variety of publics; it can no more be confined to a moral culture of opposition than to the professional world of science and the official public sphere. The fact is that late modern society cannot be integrated by a unitary public sphere governed by the norms of consensual communication.

With respect to disputes over biotechnology, the implications of this discursive and pluralist approach is that the aim is the mobilisation of critical communities for creating participative public spheres geared to defining problems and proposing solutions. As Moser (1995, p. 15) argues, the 'question is how to make the politics of knowledge a subject of investigation, and how to integrate such a politics into discussions of biotechnology.' Biotechnology touches on the core of the political today in that it opens up new discursive spaces in society. These spaces are centrally filled by a politics of societal responsibility since what is being threatened is society itself. Biopolitics is conducted in the new public sphere associated with the mass media in which a plurality of social actors collide over definitions of risk. The new politics is then ultimately a politics of knowledge; one that centrally resolves around the politicisation of science whose neutrality has collapsed.

This is perfectly encapsulated in the BSE controversy, which since Chernobyl has been the most consequential debate on science and society. It exemplifies how the organised irresponsibility has undermined the ideology of the objectivity and neutrality of science. To take Beck's (1995a, pp. 128–9) evocative metaphor, politics is no longer about the redistribution of the social product, for the cake has been poisoned. The dispute over the cake's toxicity brings other areas of power into focus, which Beck calls the 'relations of definition': what are the rules by which risk is judged? Who is the expert? What is to count as proof? What is the appropriate compensation? In the case of BSE disputes over risk have spilled over into a crisis in the national identity of Britain – the poisoning of a powerful cultural symbol – and the impotence of parliamentary sovereignty. The food industry, one of the core areas of biotechnology, has been politicised in the recognition that science has created a system of production in which agriculture, science and technology, and the state are locked together in a complex web of relations. The BSE debate is primarily a debate about societal responsibility and raises major issues about social agency. In the context of discursive democracy, it reveals how biopolitics, which is rapidly becoming the new ecological masterframe, is primarily concerned with the mobilisation of critical public opinion in the struggle to break science's monopoly of scientific discourse.

An alternative to Habermas's position, which I have argued has not satisfactorily resolved the relationship between discursive democracy and a dialogical model of communication, would be to link the idea of discourse to societal responsibility. In this context the ideas of Hans Jonas provide the most useful point of departure, but suffer from the defect that his conception of a new global ethic of responsibility is too Platonic in that it wishes to retain a constitutive link between science and knowledge. While Jonas is ultimately concerned with ontology, Habermas's approach offers the possibility of grounding responsibility in communication. His model of discourse suggests a notion of societal responsibility as a macro-ethics. Societal responsibility is capable of compensating for the deficiency in what Apel calls the 'western complementarity thesis' by which ethics (moral rationality) and politics (instrumental and scientific rationality) are separated. The notion of societal responsibility advocated here differs from conservative ideas of social responsibility (the responsibilities the dutiful citizen has to society) and also parts company from the essentialistic 'biopolitics' of much of contemporary social movements. Societal responsibility, as a new discursive paradigm, links politics and ethics in a macroethics capable of challenging both the moral neutrality of science and the oppositional politics of new social movements.

Tying this in with contemporary biopolitical debates on biotechnology, we can see how science's monopoly on scientific rationality has been broken with the opening up of new discursive spheres in society. In the definition of biotechnological risk and the appropriate social response, the mantle of science is now being claimed by many social actors – governments, EU policy-makers, the media, farming lobbies, the public – whose discourses have been scientised. These new discourses point to a transformation in our conception of democracy along participatory lines, though the actual form that participation will take is unclear since the conflicting discourses are based on quite different interests. But what can be established is that one major theme in these discourses is the question of societal responsibility. Therefore a crucial question for research and policy-making in the future will be to determine exactly how biopolitical debates revolving around biotechnology could be capable of institutionalising a democratically constituted ethic of societal responsibility.

3 Biotechnology as Expertise
Barry Barnes

The relationship between ordinary members of society and recognised expert practitioners is of great sociological interest, and is the subject of an extensive literature. My purpose in this chapter is to look at the relationship in the case of biotechnological expertise, and to point out some of the difficulties to which it currently gives rise. However, nothing fundamental separates biotechnology from any other form of expertise. Biotechnology is a mundane and long-established form of technology, as any Guinness drinker will be well aware, and it would be seriously misleading to treat it as possessing unique characteristics. Accordingly, it is sensible to begin with some completely general remarks about experts and their social credibility, remarks which will be relevant to biotechnology but not only to biotechnology.

GENERAL FEATURES OF EXPERTISE AND ITS CREDIBILITY

Expert roles imply the existence of a division of intellectual labour. Some skills and forms of knowledge are carried by the occupants of specialised roles or occupations, on behalf of the rest of us as it were. Because the specialists carry the knowledge and skill in question, the rest of us need not. A gain in efficiency is the presumed result. But the price of the efficiency gain is a degree of dependence. The ordinary member has to trust the expert, both as bearer and user of knowledge, and to draw conclusions on the validity of expert knowledge from the recognised standing of the particular expert in question and the expert profession to which she belongs. Where there is no trust then *ipso facto* there is no expertise. Where there is no trust all *prima facie* expert pronouncements must be subject to detailed evaluation by the ordinary member, which implies an impossible equivalence of knowledge and competence between expert and ordinary member, that would result in no efficiency gain via specialisation. Experts just are trusted specialists who carry a differentiated body of knowledge and competence on behalf of those who trust them and accord them the standing of experts by way of signifying that trust.

Simply by offering this definition, however, important ethical and practical issues are evoked. Knowledge is power. By virtue of being trusted as a source of knowledge experts are invested with power. How then are those who depend upon and trust expertise to keep its power within bounds? And more generally, how are societies which are so strongly hierarchical at the level of knowledge and skill to remain nonetheless genuinely democratic and able to sustain authentic political equality? Jürgen Habermas is perhaps the best known of the many writers to have addressed these issues. He takes what is, in the last analysis, an optimistic view of the relationship between the laity and expert professionals. True, expertise is currently unduly dominant, and may long remain so, but there are means of controlling expertise and keeping it in its appropriate place. What is required is strong 'reflexive' interactions between lay and expert forms of culture; ways of ensuring that such interactions are conducted rationally in conditions free of coercion and constraints on the flow of information and argument; and a recognition that the consensus which is generated by such rational interaction is the appropriate basis for social action.[1]

Habermas's writing has become an enduring point of reference whenever ethical problems of expertise are discussed, but his ideas have nonetheless attracted criticism, and indeed they merit it. In the present context, for example, it is arguable that Habermas evades the crux of the problem represented by expertise, and offers, not a genuine solution to it, but mere palliatives designed to minimise its consequences. An interplay between laity and experts sufficient to solve the problem of imbalance of power would be an interplay which eliminated the imbalance of knowledge as well, and hence expertise itself. Expert–lay interactions are, almost by definition, interactions lacking in the full and free communication asserted as necessary by Habermas. Yet Habermas recognises the necessity of division of intellectual labour and the essential role of expertise in modern societies.

Habermas's work can also be criticised for encouraging the belief that expertise is a unity which can be contrasted with another unity, the lay public. Empirical study suggests that expertise is fragmented, and that experts are more likely to be found engaged in conflict with each other than with ordinary members. Particularly worth citing here is the work of Nelkin (1975, 1992), who has studied a wide range of technical controversies associated with important policy decisions. According to Nelkin (1975), most instances where scientific experts are introduced into debates about controversial policy issues reveal the following characteristics. First, patrons seek experts to legitimise their existing plans and to use their command of technical knowledge to justify setting aside opposed arguments. Second, while expert advice can sometimes help to clarify technical constraints, it also is likely to increase conflict, since both sides will employ it and expert controversy will arise around the initial policy debate. Third, the extent to which technical advice is accepted depends less on its validity and the competence of the expert than on the extent to which it reinforces the existing positions of different audiences. Fourth, those opposed to a policy decision involving change need not assemble the same amount of evidence and expert support as those pressing for it. Fifth, the conflict which typically arises amongst experts on contested policy matters reduces their political impact and importance. Finally, the role of experts in these cases appears to be as described whether the experts are from 'hard' or 'soft' sciences.

There is indeed a rapidly increasing general awareness of the ubiquity of expert controversy in modern society and of the disturbing extent to which experts have lost their independence to patrons and employers. The kinds of case that Nelkin has studied in detail are now being aired in the media. The continuing saga associated with BSE is a spectacular example, wherein lay spectators will not have failed to note interestingly different 'official' expert constructions of risks which correlate strongly with nationality.[2]

The fragmented character of expertise can also be revealed by internal studies. These indicate the existence of many and various expert cultures, each of which exists as an inherited tradition of knowledge and competence with a particular ancestry. Expert practitioners acquire their competence from previous practitioners in a specific tradition, accept it on the authority of those practitioners, make use of it (and thereby develop and change it) and pass it on as an authoritative source of expertise to their successors. The credibility of a given body of expertise is thus the credibility of a specific tradition which not even experts themselves evaluate individually. Individual experts trust the ancestors (and each other), just as they themselves are trusted by lay persons. Bodies of expertise are not merely externally constituted and sustained by trust, they are internally so constituted and sustained as well. Expertise is never trust-independent and never tradition-independent. And hence it is not the case that trust-independent processes of rational argument and impartial monitoring of the real world will automatically suffice to correct and improve bodies of expert knowledge so that they fuse together into one thing – a single coherent unified orientation to that world.

Different traditions of expertise permit different constructions to be put upon reality: different experts may orient to and order reality in different ways. Nothing enforces and guarantees agreement between experts either in their overall accounts of things or in particular cases wherein expertise is employed. Reality itself, needless to say, is indifferent to how it is oriented to or described, and never voices any dissatisfaction with whatever specific construction is put upon it. Hence to accord credibility to a particular body of expertise, or to a particular account or practice within it, may often be a matter of cleaving to it in preference to some alternative. And such a preference is likely to reflect not any alleged degree of correspondence with reality of the favoured expertise but rather contingent, sociologically interesting factors like the perceived authority of one tradition of expertise as opposed to another, or the perceived extent of consensus amongst the expert carriers of the tradition, or the perceived disinterest of the experts in question concerning the issue on which they are pronouncing. This of course means that different kinds of expert advice and expert intervention are likely to be acceptable to different audiences, and that controversy must be expected to be endemic between opposed bodies of expertise and their related audiences, just as Nelkin has documented.

THE CREDIBILITY OF BIOTECHNOLOGICAL EXPERTISE

Expert professionals in biotechnology should recognise that Nelkin's conclusions are relevant to them. They should expect their practices and pronouncements to be more or less attractive to different audiences, and to have the credibility of their claims opposed as a matter of routine. Rather than expecting to be able to dispose of such opposition purely by means of logical argument and references to evidence, they should recognise that what they might propose as conclusive argument or decisive evidence will not necessarily be accepted as conclusive or decisive, and indeed will not be indefeasibly conclusive or decisive. They should not be surprised if they are opposed not just by lay audiences and interest groups, but by conflicting expertise mobilised by those audiences and interest groups, including conflicting biotechnological expertise. In summary, they should avoid the potentially costly misconception that 'having the truth' or 'being right' is bound to lead to victory in controversy, as though truth and rightness are forces of nature which somehow act upon other minds to enforce assent.

All this is merely a summary of mundane findings about the credibility of any form of technical expertise. There is, however, one aspect of the problem of expert credibility that, whilst being by no means specific to biotechnology, arises with peculiar intensity in relation to it. This is the question of how far an issue is properly designated as a technical and/or empirical one, and how far it transcends the realm of the technical/ empirical altogether. A number of issues where this question arises have come to surround new reproductive technologies and aspects of genetic engineering.

It may be thought that the general sphere of competence of technical experts is simple to delineate: they have standing on what is the case and what is not, what works and what does not. Theirs is the realm of empirical problems but not moral ones, of 'is' but not 'ought'. In our different-iated societies, technical expertise has separated out as part of the division of labour to focus wholly upon instrumental issues. Evaluative problems are left to other kinds of specialists, or else remain wholly in the sphere of everyday common sense understanding. This, however, is far too simple a picture.

Let us look back a century or more to a period when expertise was less differentiated and specialised. In the early nineteenth century, priests and clerics, besides claiming special standing on moral and religious matters, thought fit also to pronounce upon matters of fact. Indeed, they did not always mark a distinction between moral and factual matters. Only slowly, under pressure from growing numbers of professional scientists and scientific experts, did they learn to leave what became defined as a sep-arate empirical realm to others and content themselves with attempting to monopolise 'moral' and 'religious' contexts. Scientists and associated experts became recognised as having standing in the profane realm; priests and clerics claimed standing in the realm of the sacred (Turner, 1974, 1978).

Thus emerged what ever since has been a universally acknowledged boundary between the sacred and the profane in the realm of expertise, albeit a constantly changing one. Over time the tendency has been for the realm of the sacred to shrink and the profane realm of the technical to grow. Increasingly, the sphere of technical expertise has extended itself until it has made practically the whole of the non-human realm its own, and has indeed made substantial incursions into the human realm as well. Indeed there are fields of technical expertise which treat the entire human body simply as a profane object and other fields which extend that orienta-tion to cover what are commonly called mental phenomena as well. Nonetheless, the boundary remains, even though it is constantly being redefined and is drawn differently by different cultures and sub-cultures, even different expert sub-cultures, within our society. And phenomena which blur the clarity of the boundary provoke disquiet and anxiety, just as confusions of the sacred and profane tend to do in all kinds of societies: this is evident enough in our orientation to corpses, foetuses and embryos, excrement, and so forth.

Since the boundary is culturally variable and constantly being redefined and reconstituted the particular problems it creates will vary and the anxiety-provoking borderline entities associated with it will change. But the boundary itself is bound to persist and cannot be regarded as something that will eventually disappear when a thoroughgoing rationalisation of our social and cognitive practices is completed. There will never be a culture wholly and exclusively concerned with means and instrumentalities; for reference must always be made to the (human) entities for which something is a means, in relation to which it has instrumental value. Once the differentiation of means and ends is made, there must always be some analogue of our present contrast of persons, who have ends, and things or processes which count as means to these ends. Thus there will always be some analogue of our present contrast of a sacred realm of persons/souls, living entities, active agents with responsibility, conceived as indivisible wholes, deserving of respect; and things/objects, inanimate entities, causally moved, conceived as divisible and manipulable, and addressed purely instrumentally as means. And even if the human body is subsumed entirely into the latter framework the human being will not be.

Whereas most expert professions are perceived as located wholly in the profane realm of the instrumental/technical, some of them straddle the sacred/profane boundary. Medicine has long occupied such a position. The medical profession has claimed standing on both empirical and moral matters and has been oriented to human bodies both as material objects and as sacred entities: its members have always acted as technical advisors on the well-being of the body, but they have also possessed powers to regulate bodily movements and have often exercised them over and against the inclinations of those who inhabited the bodies. And whilst it is perhaps remarkable how readily and meekly bodies have been yielded up to the control of the medical profession, it is no less true that this control has been enforced when necessary even against very considerable resistance. This serves as testimony to the traditional standing of the medical practitioner as a specialised moral agent and to that of the body with which she deals as a sacred entity not invariably available as a means to individual ends even where the ends are those of the person who inhabits the body in question.

THE SCIENTISATION OF MEDICINE

Medicine straddles the boundary between the sacred and the profane, which is a fluid boundary. The persistent expansion of the profane realm expresses itself in this context as the scientisation of medicine, with consequent changes in the nature of the expert role. Formerly, doctors and physicians were healers, restorers of health, where 'health' was at once an empirically applicable and an evaluative concept reflecting a role both technical and moral. Increasingly, however, their role is being reconceptualised in ever more technical terms: they are becoming technologists of the body (and even in some cases of the mind), and medicine is thus becoming a branch of technology – of biotechnology, to be more specific. This is a profound, ongoing change, some of the profundity of which can be obscured by the continuing euphemistic use of the inherited vocabulary of the profession, but which may readily be illustrated by means of examples.

Let us take just two of many striking illustrations. First, consider the medical use of the new reproductive technologies, IVF procedures for example. These procedures are monopolised (very profitably) by the medical profession and employed in what is routinely referred to as the 'treatment of infertility'. This invites an analogy with routine cases of the treatment of pathological conditions like food poisoning or pneumonia, but little reflection is required to perceive how misleading this is. IVF is a technological fix for the condition of childlessness in a specific social context, as often as not where the childlessness of the 'patient' is a consequence not of bodily pathology but of custom. The 'treatment' may involve painful invasive procedures enacted on the body of a perfectly healthy woman (Crowe, 1990); it is nothing to do with health and healing.

A second, substantially equivalent example is provided by euthanasia. Let us completely set aside the matter of its desirability, and note merely the presumption, of both professionals and public, that if and when euthanasia is introduced it will be the responsibility of medical practitioners. Indeed in some countries practitioners are already volunteering to accept (and monopolise) the task of accelerating death under this rubric. Again we have a move towards a conception of medical expertise as technique available for whatever purpose, and away from the conception of medicine as concerned with the virtuous state of health – although given the fluidity of language and the ingenuity of its users we should not be surprised if in due course we start to hear talk of death as a healthy state.

What we see illustrated by these examples is a tendency for medicine to become redefined as a form of purely technical expertise, and for the human body to become a profane object for manipulation by its 'owner'.[3]

For some, this may be an altogether desirable development, which permits a strong division of technical labour but keeps moral and evaluative issues and the maintenance of our shared sense of sacredness wholly within the province of the common culture and the democratic discourses of every-day life.[4] But at the same time the euphemistic vocabulary deployed in the context of the examples stands as testimony to the unease generated by such changes, and as an indication that they are always liable to be contested.

IVF, most of us might think, is not well described in the traditional evaluative language of medicine as a treatment for a pathological affliction. But to describe it as a set of techniques to avoid childlessness in a socially and psychologically tolerable way – techniques to be evaluated purely as means, in terms of cost and efficacy – is not bound to prove more acceptable. Thus, some might seek instead to render some of these techniques as violations of women's bodies. Others might characterise them as intrinsically immoral practices by expert and client alike, involv-ing the casual manipulation of sacred materials (eggs, embryos). Others again might point out that to speak of 'technique' is as unsatisfactory as to speak of 'treatment', since in reality there is an evolving practice here best understood as 'research'. And indeed there is some professional recogni-tion that IVF 'treatment' is invariably at the same time 'research', and much of the UK legislation following the Warnock Report (1984) emerged precisely from the need to define orientations to those 'spare' embryos made available for research by surgical intrusion into super-ovulated women being, as is said, 'treated for infertility'.

IVF is an example of an issue which not merely engenders conflicting technical assessments but also conflicting conceptions of what kind of an issue it is and whether or not it transcends technical considerations alto-gether. Nor is it enough to say that as well as technical problems relevant to the issue there will be other problems as well, transcending any purely technical approach. Nor will talk of an issue with both technical and evaluative dimensions necessarily help to settle matters. These are just two of many possible contestable constructions. Another is that there is an irre-ducibly evaluative problem here, one that needs to be resolved wholly in the context of an undifferentiated everyday discourse wherein technical experts have no special standing. Just as there is no indefeasible method for establishing the superiority of one expert technical analysis over another conflicting one, so there is no method for determining the best account of where a purely technical analysis is appropriate and where it is not. Needless to say, however, just as it is often crucially important with the former problem to move to a decision in practice so it is with the latter.

DETERMINING THE LIMITS OF THE TECHNICAL

I have said that controversies involving biotechnological expertise may involve chronic conflict not just between incompatible technical analyses but also over how far technical analysis is appropriate. In one sense, the point is generally recognised. Most of us would accept that some things may rightly be treated purely as means to ends whereas others may not be. Scarcely anybody would wish to argue that a purely technical utilitarian analysis is properly applicable everywhere, so that funereal practices might appropriately be regarded as a form of waste disposal, or necrophilia addressed in terms of a principle of maximisation of happiness. And in the light of such examples it is easy to see the need to agree on a demarcation of the limits of the technical, and how controversy might arise in the course of attempting to do so. What is less easy to see is why the strategy of demarcation itself is a potentially controversial one, and why the claim is sometimes made that there are no 'purely technical' issues.

Habermas (1971) has noted how undesirable consequences may follow a strong separation of the technical and the moral. The latter area may then be dismissed as 'subjective' and 'irrational', and debates decided wholly in terms of pseudo-technical arguments about what is feasible. Or the way may be opened for a diffuse, shallow utilitarianism which 'solves' the 'moral' side of a problem by reference to whatever individual ends or goals are found immediately adjacent to it.

Again, once the existence of a domain of wholly technical interest is agreed, it is liable to become highly fragmented and atomised, as the specialised competences of technical experts are used to analyse it. The environment may come to suffer insult from hundreds of industrial processes, all reckoned to have 'negligible' polluting side-effects in hundreds of separate public enquiries. The human body may suffer insult from endless substances, all separately investigated and tested and pronounced 'safe'. Some of those who press for a healthy existence in an unpolluted environment do rightly wonder whether they will not risk defeat by salami tactics, once they allow their vision to be translated into technical questions about particular decisions.

It is, moreover, notoriously difficult to handle hypothetical possibilities in the context of technical debate, yet such possibilities often constitute much of the case made against biotechnological innovations and are surely the genuine basis of many anxieties and reservations. Thus, a part of the opposition to IVF rests upon technical developments in this area opening the possibility both of a greater domination of women by men and/or medics, and of the growth of explicitly eugenic practices in egg selection and so forth; but in fragmented technical debate on immediate particular issues these longer-term general possibilities tend to be discounted.

Evidently, there are many dangers involved in identifying a specific domain, however small, for monopolisation by technical experts and those in a position to support to employ them. But this is not the only reason for reluctance to allow the demarcation of such a domain. If the limits of technical expertise are to be clearly drawn, then a method of delineating them is required, and this itself may become a contested matter. Moreover, the problem of the proper scope of technical expertise may recur in the course of the contest. Consider, for example, the problem of for how long human embryos should be made available as material for scientific research. In Britain, legislation currently permits experimentation for up to 14 days after the initial collection of an egg. If we might put it so, at the end of the 14th day the 'embryo' makes a transition from being a profane object, manipulable/researchable in any way, to being a sacred object not available for research but deserving of more respectful treatment.

The 14th day legislation is generally acknowledged to be a compromise between researchers, some of whom feel unduly restricted by the time limit and would have wished to remain oriented to embryos as mere inert material for a much longer period, and those with a respectful orientation to embryos. But it is a specific kind of compromise. The Warnock Committee, which inspired the legislation, was insistent upon setting an objective marker of the limits of the technical. Thus, it was unwilling to give extensive consideration to claims that research should cease at the point where embryos might feel pain; for such feelings are invisible and conjectural. Something verifiable, even if arbitrary, was preferable to something unverifiable, even if meaningful.

This emphasis on what is publicly verifiable is, of course, merely a reflection of the form that any legislation must have in societies marked by an ever-increasing scientisation of medicine, rationalisation of law and democratisation of politics. In these societies the marking of the limits of the technical must itself be a technical task. Thus, there is a sense in which the current state of English law, apparently a compromise between opposed perspectives and interests, is in truth a defeat for that interest which insists that the embryo is sacred and deserving of respect, precisely because the 'compromise' is expressed as an arbitrary convention defined by the clock, the universally recognised symbol of concern with prediction and technical control. It is worth reflecting as well that the concept of compromise itself belongs, as it were, in the realm of profane discourses, not sacred ones.

Even in this social context, however, the preference remains to ground law on more than 'mere' convention, even if it is verifiable convention. It is interesting to note how, following the proposal of the 14-day rule, suggestions began to appear that it corresponded to a 'real' natural boundary, a physical discontinuity in the development of the 'embryo' itself (Crowe, 1990). References were made to the appearance of a 'primitive streak' at about 14 days, a physically differentiated feature which was argued to mark the transition from the 'pre-embryo', an unstructured, undifferentiated, profane, researchable object, to the 'embryo proper'. Questions were later raised about the visibility and 'reality' of the 'primitive streak' and about whether it might have been invented in order to legitimate as 'natural' what was previously merely a 'conventional' boundary. But the interesting questions are not so much whether the streak is real and the pre-embryo exists, as whether a distinction was here being marked out of 'political' rather than 'technical' expediency and, if so, why it was that 'natural' discontinuities should have been thought easier to defend as the bases for demarcations than 'mere' conventions.[5]

In a sense of course it is 'only natural' that we should wish to correlate discontinuities in practice and attitude with discontinuities in the physical world, and that where the world is less than helpful in providing discontinuities anxieties arise. Thus, it is frequently said today that death is a process, which may on occasion be drawn out and lacking any clear beginning or end, but this may not be an attractive point of view for the distraught relatives of a comatose individual on a life-support machine. If it cannot be strictly connected with some 'objective' marker, then switching off the machine becomes specifically a matter of individual responsible action, where the responsibility may often be unwanted or even intolerable.[6] One solution here has been for an expert to determine 'point of death' through means incapable of explicit description: trust in such an expert may replace trust in signs given by the world itself as the basis for confidence in the objectivity of a judgement. But the experts who have been trusted in this way in the past, and allowed discretion to define 'what is really happening', have been perceived as both morally and technically trustworthy. If expertise continues to be reconceptualised as more and more a purely technical matter, then the increasingly limited extent to which experts are trusted will inhibit them from operating in this way. Without trust in practitioners as moral agents, it becomes harder, for example, to be sure that the judgement of decease in ward A is not associated with the need for an organ transplant in ward B. Hence the demand for visible markers is likely to grow, and, perversely, for the technical expertise needed to identify them, precisely because of the narrow scope of the trust accorded to purely technical expertise.

CONTESTING THE AUTONOMY OF THE TECHNICAL

There will always be some groups or interests which object to the definition of certain issues, particularly perhaps issues involving biotechnology, as purely technical ones, or even as issues where technical questions must first be determined in order to resolve them. On the other hand, the forums in which these issues are currently debated invariably give prominence to technical issues and often take it for granted that the key to securing consensus on a decision is determining what technical evidence is to be believed. This is the case even where forums actively encourage the involvement of protest movements and representatives of deviant perspectives. Thus, participants in these forums sometimes find issues being framed in unacceptable ways, and feel obliged to contest the framing rather than to offer evidence of a technical kind within the given frame. The 'failures' frequently reported in settings like science courts or technology assessment colloquia may often be the consequence of orientations of this kind: official presentations of such settings as opportunities to evaluate information do not preclude their reconceptualisation as theatre.

Groups and movements that ignore the presumed framework for debate in contexts such as this are sometimes held to be irrational, which term is widely used to describe challenges to the extension of a wholly technical, utilitarian approach to problem-solving. Needless to say, when utilitarians and believers in the limitless potency of the technological fix taint those who oppose them with 'irrationality' they rarely seek to back their rhetoric with reasoned arguments, and indeed it is hard to see where they could find any.[7] What would actually be irrational for these oppositional movements would be for them to acquiesce in a frame for participation which celebrates the very thing they challenge, the priority and independence of a realm of expert technical discourse.

Those who fear the constant extension of the realm of the technical are institutionally disadvantaged in current circumstances. Arrangements designed to be fair to both sides in a technical controversy do no favours to those who would deny that the issues being discussed are technical ones. Moreover, the institutional disadvantage is compounded by cultural disadvantage. To seek to roll back proliferating technical/utilitarian conceptions of problems is to attempt to further processes of de-differentiation and resacralisation in a culture where language and symbolism have already become very highly differentiated, where vast areas of discourse and activity have become thoroughly profane, and where both kinds of change have widespread acceptance as rational and progressive. In such a culture the project is hard to justify and rationalise. Thus, for example, the attempt to resacralise embryos and de-differentiate our orientations to them so that they are never in any situation treated merely as materials for manipulation involves advocating something very like the creation of a taboo. And whilst anthropologists offer interesting accounts of the value of taboos in undifferentiated societies, they are not things that can be argued for: to engage in argument for them – given current conceptions of what counts as argument – is precisely to bring them back into the means/end frame from which, of their nature as it were, they must remain aloof.

It would be wrong, of course, to assign all of the more radical and un-orthodox forms of resistance to the expanding powers of technical expertise to the above kinds of consideration. No doubt it will be just one of several factors in any given case. However, it is a factor which is easily underestimated, not least because those groups and movements alienated by technical approaches often find themselves obliged to adopt such an approach themselves, for the sake of their own credibility. Technical expertise is now an indispensable weapon of those engaged in resisting technical expertise. Indeed, the most successful opposition to the purely instrumental approach of industries and bureaucracies to the environment, and to human beings, seems to be mounted precisely by movements which adopt that approach themselves. PR, media manipulation, image management, the propagation of useful untruths, are now all routine features of the 'if you can't beat them, join them' strategy of successful oppositional movements. Greenpeace is an obvious example here, and its recent campaign against the disposal of the Brent Spar oil storage structure a nice illustration of its effective operation.

'Effective operation' of this kind, however, risks complete assimilation into the system of instrumentalities that movements seek to oppose. It abandons any attempt to oppose differentiation, desacralisation and instrumentality *per se*. Yet we need to give space and sustenance to those who genuinely seek to oppose these processes, and to make sure that their voices are heard. For the perversions to which extreme developments of these processes may lead have long been clear and evident, not just through sociological studies like those of Max Weber but even more perhaps through the imagination of creative writers like Aldous Huxley. And if a consistent and uncompromising utilitarianism is repugnant, then how is it that an inconsistent and partially expressed one should be assumed to be beyond all criticism?

Provision for the contestation of what is technical must be made in modern polities, just as it is already made for contestation within the realm of the technical itself. But to make provision of this kind will not be easy. It is the need to make allowance for the evolution and contestation of different forms of culture that is being spoken of here; and it is rich, cross-cutting, inclusive interaction within the life of entire societies which meets this need, rather than the special forums designated for the evaluation of the competing knowledge-claims of technical experts.

The writer who has dealt most extensively with processes of differentiation and rationalisation in modern societies is Jürgen Habermas. His account of these processes has evolved into a remarkably conservative one. Our present state is said to be one where economic, productive activity is very highly differentiated and rationalised, but where a 'lifeworld sphere', encompassing familial and informal social relationships, is relatively less so. This is very much as it should be: differentiation and rationalisation are good in the context of the economy, but would be disastrously dysfunctional in the context of the lifeworld. The need is to resist the 'colonisation of the lifeworld' by the kinds of action orientation valuable and appropriate only in the sphere of the economy and the polity.[8] Clearly, this evaluation of the proper domain of the technical/utilitarian can be contested: both radicals and reactionaries might be tempted to ask why our present condition is so like that of Baby Bear's porridge.

However, whilst Habermas's evaluations are problematic and contestable, the implied description of our present condition is a plausible one which is indeed very widely recognised and accepted. In comparison with other areas of modern societies, activities in the context of the family remain relatively undifferentiated and multifunctional: there has not been extensive rationalisation and specialisation in this context and few actions have become oriented solely to specific technical/utilitarian concerns. At the same time, however, the means to effect such changes are becoming available, pressures to make use of them are growing, and use is increasingly being made of them. Familial and sexual relations are indeed 'threatened' with differentiation and rationalisation. They stand today, perhaps, much as work stood a couple of centuries ago, and to imagine a scenario of possible future change for them one has only to reflect on how work has changed over those centuries. But whereas it was a developing mechanical technology that was associated with the differentiation and rationalisation of what we now call work, it is a developing human biotechnology which is making these further changes not merely conceivable but realisable.

Biotechnological innovations will in due course permit a number of presently undifferentiated concerns to be addressed and evaluated separately as so many distinct technical tasks; these may include genetic mapping and selection including sex selection, the fusing of egg and sperm, the management and growth of embryo and foetus independent of the parent's body, the social structuring of childrearing, the involvement in sexual activity purely as pleasurable recreation. Many existing social roles and institutions will be altered out of recognition if even just some of these possibilities become actual. Because of this, human biotechnology, and by association biotechnology generally, will inevitably be perceived as specially placed in relation to some of the major long-term tendencies of social change. And this in turn will make the issues I have been describing of particular import in relation to it, and worthy of the special attention of its expert practitioners.

ACKNOWLEDGEMENTS

The reflections and criticisms of members of the original workshop at Cork, and the work of the members of the Centre for European Social Research have helped me to improve this paper, as have the comments of my colleague at Exeter University, Professor Bob Snowdon. I also want to pull from the bibliography the interesting work of Christine Crowe (1990), and record how useful I found her discussion.

Part II

Constructing Values: Public Communication on Biotechnology

4 Shifting Debates on New Reproductive Technology: Implications for Public Discourse in Ireland

Orla McDonnell

INTRODUCTION

This chapter maps the public construction of new reproductive technology (NRT) in relation to international debates and the implications of those debates for the construction of a public discourse in Ireland. It sets out to reconstruct the social field of NRT as it is discursively constituted at the level of public discourse: in other words, how NRT is shaped and reshaped by particular symbolic configurations. The social field of reproduction within which NRT practices are symbolically negotiated is configured by different social forces and institutional contexts – the state, the media, the family, feminists, professional bodies, individual patients and doctors, etc. This social field is dynamically configured in struggles over social meaning and legitimacy. The chapter's limited methodological focus on public discourse as it is constituted through popular media does not testify to the various micro-processes involved in the struggle over meaning, but it does indicate how new subjects and objects of discourse are constituted and raises questions about how public debates are mediated by particular configurations of a shifting cultural landscape.

Over the last two decades developments in biomedical technology and, in particular, the rapid rate of technological developments in reproductive medical practices have raised public disquiet about the role of technology and science in mediating interpersonal and social relations. The transgression of the symbolic order between culture and nature is the dominant operative framework in which the ethical, moral and social dilemmas of NRT are concretised. However, as this chapter sets out to argue, the symbolic order is culturally constructed and this is reflected in the shifting debates on the implications of NRT. For the purpose of locating the social evaluation of NRT and its cultural diffusion Van Dyck's (1995) reconstruction of the genealogy of NRT as an international public issue is invaluable. International events and their particular narrative structures not only permeate national discourses, but their cultural resonance provides insight into wider socio-cultural considerations. This is not to suggest that the meanings attributed to events elsewhere in themselves become an evaluative tool for assessing the peculiar or local, either in terms of 'lessons to be learnt' or endowing particular cultural models with a linear and transferable quality. Instead, the relevance of Van Dyck's reconstruction of the international dimension of the debate to the Irish context enables us to contextualise particular meaning systems within broader structures of referentiality. The analysis of public discourse structures through their context traces the process of 'normalisation' invoked in the NRT debate and helps us to make sense of how this process is constituted and mobilised by different actors.

SITUATING THE 'PUBLIC' DEBATE ON NEW REPRODUCTIVE TECHNOLOGIES IN IRELAND

The need for a legislative response to the rapidly developing field of NRT is documented in a report published by the Irish Department of Health in 1986. The context for a proposed political mandate was the proliferation of international public debates on the risks of NRT. However, since the late 1980s the preferred institutionalised response has been to defer public debate and to leave policy frameworks to the self-regulatory activity of the medical profession. Elsewhere, particularly in Nordic and central European countries, the dominant institutional response from the mid-1980s has been to set up national ethics committees as a conduit for public debate and as a precursor for the framing of regulatory politics. Such institutional responses are indicative both of the widening social competence of the state and of the emergence of 'discourse publics' (Fraser, 1989) who claim to have a stake in bioethical issues. These discourse publics become visible and are mobilised through such quasi-legal and ethical institutional contexts. A Council of Europe report comments that the institutionalisation of these types of discursive arenas is testament to 'the importance of the ethical phenomenon, the urgency of the problems raised by bio-medical progress, and the growing demand from the scientific and medical community, the general public and also the public authorities, for an effective guidance system and a *national debate on ethics*' (emphasis mine) (1993, p. 3).

While the concept of a 'national debate on ethics' has become part of the institutional imagination in certain countries over how competing claims about reproductive needs and NRT are mobilised, this process of politicisation has been less evident in Ireland. NRT lies outside the official political economy to the extent that it is unregulated by the legislature and as a contested issue with extensive ramifications it is not publicly 'debated across a wide range of discourse publics' (Fraser, 1989, p. 166). Notwithstanding, competing claims and interpretations of 'the problem' are diffused across relatively fragmented and specialised publics. Where contestation can begin to be seen to disrupt the apparent privatised contexts of sexual/domestic relationships and the clinical setting of the doctor/patient relationship it is structurally bounded within an expert discourse on needs. In other words, the experience of infertility is ordered within the proliferation of specialised psycho-sexual and medical discourses.

The absence of political or legislative disputes on NRT has meant that public contestation and, hence, public debate have been absent in the Irish context. However, as I will proceed to argue, international public discursive events have become an important reference for mobilising public concern in Ireland. Heightened reflexivity on women's reproductive health issues, and increasingly men's reproductive health, as well as the movement away from a paternalistic system of medical care have also facilitated new discourse publics outside of the specialised and professional interest spheres of medicine.

Setting the Context for NRT in Ireland

The first controversial fertility technology introduced into Ireland was Artificial Insemination by Donor (AID) in 1982. However, as I will argue in the analysis to follow, the legal and ethical ambiguities it raised were contained within a normative discourse which did not threaten a public crisis. Gamete Intra Fallopian Transfer (GIFT) was introduced in 1986 and In-vitro Fertilisation/Embryo Transfer (IVF/ET) in 1989 (Harrison et al., 1992). Taking IVF as the most controversial of these procedures, certain advantages had been reaped from the 'normalisation' of this technology: public discourse had already been played out in relation to international discursive events such as the birth of the first 'test-tube' baby, Louise Brown, the debate in Britain about commercial surrogacy and the split between genetic and gestational motherhood facilitated by expansive IVF techniques. There was also the much publicised case of the 'orphaned embryos' in Australia, which became part of the discursive debate on embryo freezing, storage and research.

These discursive events are central to the public discourse constructed through the media in Ireland. As cultural imports they carried their own scripts and at the same time provided interpretative and communicative frameworks for a fledging Irish debate, which is beginning to impact on the manner in which NRT and infertility are politicised. There are certain features which the Irish debate shares with the broader international debate, particularly the dominance of scientific interpretative frameworks. The dependence of the media on the medico-scientific community has led to the voice of science providing the dominant structure of meaning, which defines and constrains public discourse into a narrow set of clearly demarcated ethical definitions. The scientific mode of argumentation is also strategically adopted as a mobilising rhetoric in counter-discourses to the dominant scientific-medical one. Public interest, in shaping the moral, practical and legal norms through which NRT is institutionalised as a medical response to specific needs, has until recently been visibly absent in Ireland; however, we are beginning to see the emergence of new coalitions of interests. Increasingly, ethical guidelines and moral insistence have little relevance to the socio-cultural and legal complexities raised by the expanded applications of NRT. This has been highlighted more recently in relation to the ethical and legal dilemmas thrown up by the thousands of frozen embryos in British laboratories and the widely publicised case of the British fertility patient, Mandy Allwood, who became pregnant with octuplets.

Having set out the structural parameters and context that shaped politicisation and public debate on NRT in Ireland, I will now address the international context of that debate, providing a conceptual and normative map of the dominant narrative and discursive contingencies which impede the emergence of dialogue on the basis of negotiated meanings in this particular field of reproductive politics.

DISCOURSE SCRIPTS AND THE INTERNATIONAL DEBATE ON NRT

Van Dyck (1995) reconstructs the international debate on NRT into a typology of four phases: the construction of the 'need' for IVF constituted by a scientific paradigm which she terms the 'normalising phase' periodised between 1978 and 1984; the development of a 'feminist countermyth' (1984–7), which is recast by the referential terms of the dominant discourse and marginalised as an oppositional voice; the 'naturalising phase' (1987–91) mediated through a new politics of imagery which represents science as the dominant episteme. This later phase is also characterised by legislative frameworks which are the precursor for the development of a rights discourse in the fourth phase (1991–5). This chapter does not attempt to trace the Irish debate within the same sequential map; instead, the latter's salient scripts or narrative structures are identified as discursive devices which locate how meaning is dynamically constituted in this context, and how this meaning serves as a resource to be mobilised by different actors.

Since the first public pronouncements on the wonders of the 'test-tube baby' a complex cultural repertoire has served to normalise NRT and the role of medicine in fighting infertility as a modern pathology. The referential terms of the debate which have been dominated by a scientific rationale are encoded in expanding scripts or meta-themes, which are recast as the debate develops. The key moments in this development are marked by discursive events which communicate the emergence of new subjects and objects of NRT. The genealogy of NRT finds its source in the 'plague/ epidemic' metaphor which constructs infertility as a disease and IVF its cure in the first phase of the NRT debate (Van Dyck, 1995, pp. 77–85). This script is replayed and recast as new ethical dilemmas present themselves for public deliberation.

The script encodes a number of ideological and cultural constructs. It can be constructed as a 'punishment' or warning against social and sexual transgressions – nature's 'revenge' against women who negate their naturalistic role by deferring reproduction until it is too late: women can be both victims or perpetuators of this 'fate'. The 'victim' story appeals to a pro-natalist culture reinforcing naturalistic assertions that women cannot negate their maternal desires; these desires make women irrational in their expectations of doctors and clinicians, and obsessive in their pursuit of treatments. In the later phase of the debate we see how this narrative structure is reconfigured as the script of the narcissistic modern woman. The symbolic meaning invested in the reproductive body structures the gender sub-text of this salient narrative: women's self-identities and social identities are bound to the maternal role either as hosts, donors or mothers, and men's need to negate the contingency of their lives is bound by their investment in reproduction either as fathers, donors or 'helping' doctors and scientists. The plague/epidemic metaphor also translates biomedical arguments into a market/consumer rationale which constructs the need/demand rationale in the early phases of the debate and NRT as a right in the later stages.

As the debate becomes increasingly brokered by market and legal frameworks, Van Dyck notes that feminist oppositional discourse is recast within the market rhetoric of consumer rights and protection. This is paralleled by the popularity of the visual narrative as a public communicative device (1995, pp. 128–34). As new ethical, social and legal dilemmas present themselves in the 1990s various fragments of the above script are aligned as discursive devices to reconstitute normative boundaries. The following are examples of the emerging normative and symbolic configurations which emerge from the narrative structure of the NRT debate thus far. These exemplars are chosen because of the salient theme of the future posited by NRT developments. In other words the 'future' becomes visible as a legitimate, albeit prospective constituency of public ethics, and these examples offer an insight into the particular permeation's of a rights idiom. This future is marked by a growing demand for fertility treatments and the need for regulatory politics.

Normalising Science within a Rights Discourse

The following reconstruction of a popular script exemplifies the realignments of the dominant frames of reference which have become part of our cultural repertoire in adjudicating the expansion of rights in relation to the growing demand for fertility treatment. The cover of a *Sunday Times* magazine (11 February 1996) uses the traditional metaphor of the 'barren landscape' in a postmodernist representation of infertility. The vulnerability of nature is represented by the naked bodies of a man and woman – the collective reproductive body vulnerable to the unpredictability of nature. An arid landscape represents the dominant image of sterility coupled with the image of empty pea-pods and the shell of an egg suspended on two dead tree trunks (the fruitless tree). The egg lies outside the reproductive body, suspended in nature as an autonomous agent. The caption reads: 'The Dead Zone: is infertility a new threat to humanity?' The apocalyptic image is captured by the juxtaposition of the visual and textual narratives suggesting the context for the inside story, which is about the rising incidence of infertility. The story is contextualised by the visual essay on the science of nature depicted by the 'laparoscopic' image of the sperm entering the ovum: the female reproductive body is no longer part of our informative encounter with nature; instead it is dissipated and pushed to the back regions of our imagination by the 'clinician's gaze'. The body becomes a knowable object from the inside and this episteme has become part of our cultural understanding of infertility. The scientific 'gaze' is normalised in the construction of public debates. This process of 'naturalisation' is central to the construction of a rights discourse. The images deconstructed above reinforce the notion that science provides the 'hard facts' and the 'solutions': science becomes the dominant mode of argumentation. In appealing for infertility treatment as a 'right' the reproductive body is translated as the consuming body – the body which is in search of a social meaning which nature with all its uncertainties denies.

Redrawing Normative Boundaries

The discursive events of the birth of twins to a 59-year-old British woman at the Severino Antinori fertility clinic in Rome; the successful impregnation of a 68-year-old woman then reported to have been three months pregnant at the same clinic; a transracial impregnation of a black woman with an embryo from a white woman; and the public announcement in Britain of proposals submitted to the HFEA on the use of foetal and cadavers' ova for donation in IVF treatments – all mark the shifting normative boundaries in which these new subjects are constituted. The re-ordering of biological and social facts requires normative assumptions to be reasserted. The most salient concern to emerge is the protection against any diminution of the rights conceded to childless couples. While this right is framed in terms of the 'the philosophy of the best interest of the prospective child', cultural categories are being challenged by the conceptual shifts in corporeal boundaries associated with the splitting and commercialisation of bodily parts.

The dominant narrative constructed in relation to the 59-year-old who gave birth to twins is structured by the script of the narcissistic career woman who strategically avoids reproduction in order to establish her career opportunities and with the help of a 'maverick' scientist reconstructs a reproductive identity. This is the moral context in which we are invited to adjudicate the needs of the infertile as 'worthy' or 'unworthy'. This is part of the same script in which we are asked to make moral judgements about the inalienable market rhetoric of 'demand and supply' in the debate on the use of ova from aborted foetuses and ovarian tissue from cadavers and the controversy which has ensued in Britain about the destruction of frozen embryos in storage. The neologisms of the 'dead mother' and the 'orphaned embryos' set the normative and analytical framework in which needs are socially reassessed.

MAPPING THE IRISH DEBATE ON NRT

Asserting Normative Frameworks

Media coverage in the early 1980s of infertility treatments in Ireland was mainly concerned with AID which was the first controversial, albeit low technological treatment available. Unlike Britain, moral controversy about single and lesbian women was not amplified in the Irish press. Instead, AID was normalised by the profile given to 'married couples' which was encoded by innocuous references to the 'childless'. At the time the first AID service was opened by the Dublin Well Woman Centre (WWC) in 1982 the 'married couple' was encoded in an expansive definition of medical pathologies – 'women whose husbands are infertile or whose sperm may carry a genetic disorder ... where a woman with a Rhesus-negative blood group has already been sensitised by conceiving by a partner with Rhesus-positive blood ...' (*Irish Times*, 3 March 1982). While the availability of the service to non-married couples, lesbian and single women is not denied forthwith, it is negated by reference to the 'childless couple' which encodes a cultural assumption about the legitimacy of potential clients. The competing claims of other potential subjects of AID are 'internally dialogized' (Fraser, 1989, p. 165) in the interpretative/communicative framing of the issue. While single and lesbian women are alluded to, they do not emerge as a legitimate subject or public with distinct needs and claims. Instead, deviance is evoked to imply normality. The invisibility of lesbian and single women who may avail of the service is not only taken for granted, but it is adopted as a strategy of inverted logic to avoid open contestation over competing claims.

The need–demand rationale for the service is contextualised by the falling numbers of babies available for adoption and the restrictions imposed by adoption laws for 'some couples'. The claims of non-married and lesbian couples are structurally removed from the terms of any potential debate. The reference to adoption not only reinforces the cultural assumption that it is the legitimate childless couple who will be availing of the service, but it also serves to normalise the technology by appealing to the cultural investment in the genetic imperative of reproduction – 'the next best thing to nature'. One newspaper article frames the issue by stating that '[the WWC] point out ... that AID/AIH permits the parents to enjoy the process of pregnancy with the mother/child bonding that is absent in the case of adoption' (*Sunday Tribune*, 31 January 1981). The physiology of pregnancy, which is the central body idiom of the cultural model of the family, is evoked to naturalise the technology in terms of heterosexual relationships. Hence, the overall effect of the discourse frame is the appeal to a pronatalist culture.

In the same article the legal ambiguity of the status of the AID child is negated and normalised in terms of the assumed reproductive identity of the married couple. The status of the prospective child is guaranteed by the inevitability of the biological process and the arrangement of those facts within the context of marriage. Anonymity, therefore, is protected not by legal principle, but by what is taken to be the normative context of the family and 'father right': 'The child, when born, can be registered in the normal way and since the Well Woman Centre encourages couples to have sexual intercourse on the nights before insemination there is no way of knowing whether or not the father is in fact the husband' (ibid.). The symbolic significance of the genetic imperative further serves to normalise AID and render it acceptable within existing cultural models of reproductive identity and social relations. Another newspaper article (*Irish Times*, 5 March 1982), for example, refers to the chosen location of the sperm bank in Birmingham with its large Irish emigrant population. Explicit reference is made to the potential genetic pool for the Irish market. The only oppositional voice is that of the Catholic Church: outside of the moral theology of perverse sexuality and the sanctity of procreation the issue of incest arising from anonymous gamete donation is pursued (for example, *Irish Times*, 3 March 1982). This symbolic import has wider resonance in relation to biological ties and the social organisation of sexual relations.

The Foregrounding of Abortion Politics

Prior to the availability of IVF in Ireland the Warnock deliberations in Britain provided an intertextual context for discourses on the ethical, legal and social dilemmas raised by NRT. The Louise Brown script serves as the referential framework in which the public is invited to assess the moral implications of developments which have arisen from this 'source', including surrogacy, embryo research and embryo freezing. While public concern about the future of these developments is evoked, oppositional voices are largely confined again to the Catholic Church. The status of the embryo is foregrounded and this signposts the NRT debate in terms of its political cultural resonance for an Irish public. Feminist oppositional voices are absent from mainstream press coverage of international events in the NRT story, or as a legitimate social actor in the Irish context despite the mobilisation of the Irish Women's Movement on reproductive politics for well over a decade.

In 1983, at the height of the public debate on the forthcoming abortion referendum, IVF – prior to its provision in Ireland – is given a limited public airing. One article in a national newspaper (*The Sunday Tribune*, 21 March 1983) sets out its interpretative framework in the context of abortion politics. Oppositional voices within the medical profession are animated, not just by competing knowledge claims, but more importantly by the moral adjudication of the ethical issues involved in IVF. The term 'test-tube baby clinics' as opposed to the technical lexicon of IVF shows how the early debate is constructed around common-sense notions of science, and the household names of Louise Brown, Patrick Steptoe and Robert Edwards are used as the orienting script. The article reports on the self-imposed moratorium on IVF by the Irish medical profession. This is framed in terms of the status of the foetus. Medical consensus on what was effectively an ethical ban on IVF is normalised in terms of the dominance of a Catholic moral ethos: 'The prevalent view in Irish medical circles is that human life begins as soon as fertilisation takes place in the laboratory and a human embryo has been formed ...' (ibid.). While dissent within the medical establishment on the provision of IVF is alluded to, anonymity is afforded to the dissenting voice as a structural feature of ethical prohibitions, rather than as a political issue of censure. This consensus is framed by an identity politics which appeals to the notion of an ethos in Irish medical practice which is culturally sanctioned. The innocuous descriptions of infertility treatments available in Ireland reported in the media in the early 1980s should be read against the volatile political context of the first abortion referendum (1983) which centrally involved political conflicts within the medical profession in relation the moral, ethical and social implications of women's reproductive health.

Towards Discursive Openness: Fragments of a Debate

There is a dearth of media coverage on NRT from the mid-1980s until the media exposure of the postmenopausal story which marked the juncture of 1993/4. For the medical profession consciously to enter the public arena on the issue of NRT it has had to subvert the moral discourse of the early 1980s by appealing to a new set of scientific facts and knowledge claims. Public fears which have largely been constructed in relation to events elsewhere are addressed in terms of the ethical guidelines of the Irish Medical Council (IMC) which are assumed to pre-empt such controversies arising in the Irish context. However, the ethical restrictions which ban the use of donor gametes and the freezing and storage of embryos, and exclude non-married couples from IVF programmes have themselves become the subject of public contestation.

While the status of ethical problems continues to be framed in terms of conflicting views on the status of the embryo, the need for ethical deliberation is increasingly contextualised by the wider issues carried by the international debate. For example, in a newspaper article in 1995 the headline reads 'Questions fall on fertile new ground', with the sub-caption: 'As this century comes to a close in vitro fertilisation could become the hottest topic here in Ireland' (*Irish Independent*, 2 August 1995). The Louise Brown script has now been replaced by the script of the postmenopausal birth to highlight the shifting context for ethical and legal deliberations. The 'test-tube baby' has been replaced by the metaphor of the 'designer baby': this communicates a more sinister and hedonistic intrusion into nature, and brings what has been naturalised by appeals to a pronatalist culture into the realm of eugenic policies and constructs the notion of a 'point of no return'. The status of the embryo, however, is the salient theme and the ethical ban on embryo freezing is explicitly linked to the uncertainty of the constitutional status of the embryo. Abortion politics, although less explicitly foregrounded than in the 1980s, provides the political context in which the NRT debate is constituted. The public crisis which arose from the Australian case of the 'orphaned embryos' is amplified in order to draw a parallel between it and the public crisis which arose from the X case[1] in Ireland. While abortion politics continues to structure the emerging debate on NRT the terms of that debate have changed and a legalistic rationale has increasingly replaced the politics of moral insistence. For example, the above article asserts the need for a public debate on the ethical ban on the freezing and storage of embryos by raising doubts about the apparent security in the present ethical guidelines to preempt legal disputes over the rights of the embryo.

Within a short time of the media exposure given to the dilemmas posed by postmenopausal births, an open information day was publicly announced by the Rotunda Hospital which houses the Human Assisted Reproduction Institute (HARI) (*Irish Independent*, 15 March 1994). The entry of the medical profession into the construction of a public debate represents a key moment in the emerging process of 'discursive publicity' (Fraser, 1989, p. 167). The medical profession has increasingly become aware of the fact that IVF has caught the imagination of the public, and in the public contestation over the interpretation of infertility needs it sees itself as having a central role to play in the 'interpretation/communication nexus' (ibid.) on defining the status of the problem. The decision of the medical profession to be part of the public profile given to the issue of infertility treatments is framed in terms of its public education role, which is encoded in a therapeutic discourse.

Mobilising Issues and New Discourse Strategies

The mobilising and operative themes of a public controversy on NRT in Ireland received its main impetus from the public launch of new pressure group, The National Fertility Support and Information Group (NFSIG). The key issues highlighted by the group are the social stigma attached to infertility and the cultural closure of a medical ethos which perpetuates this stigma through a public information deficit, fear of censure within the profession itself, and the exercise of moral decisions which deny patients an adequate knowledge base in the consultation process. These grievances are publicly communicated through personal narratives which serve to mobilise the infertile as an interested public, which has been structurally invisible as a legitimate stakeholder in earlier deliberations on NRT. Since the launch of the group in late 1995 it has used a number of key mobilising stories as a means of gaining public visibility and as a mobilising device for seeking alliances with political and professional stakeholders.

The use of personal narratives has become an important discursive resource for the NFSIG in pursuing fertility treatment as a rights issue. These 'stories' are constructed to communicate the issue in terms of how the present ethical restrictions structure the everyday context of women's lives from financial and emotional difficulties to the potential threat to their health. The construction of personal narratives as a public discourse strategy has increasingly entered popular representations of NRT as a counter-hegemonic narrative. The experiential narrative is a discursive tool used to construct the 'need' for an expanding fertility service; it augments the rights idiom by appealing to common-sense notions of justice and it serves as a subjective account of a collective experience which lies behind the impetus to discourse the need. The public visibility given to these biographical accounts does not simply serve as 'public interest stories', but as an interpretative framework for negotiated meanings. The personal narrative is recast as a social commentary on the failure of the medical profession and the legislature to recognise the legitimacy of the need of an expanding community of interest.

For example, medical regimes such as chemotherapy and their implications for fertility loss have led to a growing public awareness of the status of men's reproductive health. The personal narrative of a young Irish male who deposited sperm in a British sperm bank prior to chemotherapy treatment ('Banking on fatherhood', *Irish Times*, 8 July 1996) is an account of the social processes involved in the changing status of sterility. This counter-narrative not only challenges social attitudes towards men and women's procreative identity, but also challenges what counts as legitimate medical knowledge and practice. Furthermore, it is an account of how meaning is constituted in human procreation and the exigent experiences of contemplating one's identity in a personal future which may well rule out the possibility of having genetic children of one's own. A growing constituency whose interest in a sperm bank facility is identified, i.e. men with Hodgkin's disease, leukaemia, testicular cancer, diabetes and multiple sclerosis. This relatively distinct constituency of potential 'users' whose needs are definable in terms of identifiable modern pathologies is aligned with a growing 'non-pathological' group of men whose fertility is negatively affected by environmental factors and the overall global trend in decreasing sperm counts. This broadens the definition of the problem and transforms the need idiom into a political discourse on rights, which in part is reflected by a growing network of interests publicly aligning themselves with the issue, including fertility clinics, professional bodies such as the IMC and specialised public interest groups such as the Hodgkin's United Group (HUG).

Individual experiences are collectively transformed into a mobilising issue by coopting the terms of the dominant discourse through which this need may be publicly asserted. The need for an expansive service is communicated by the 'export script'. For example, a *Sunday Times* article (18 August 1996) recounts the personal narrative of Cathy Murphy a fertility patient who had to travel outside the jurisdiction for IVF treatment involving donor sperm, and the case of Gerardine Gleeson who died in 1995 following an illness after she received fertility treatment, again outside the jurisdiction. The 'export script' which is a legacy of reproductive politics in Ireland is recast to highlight infertility as part of that political context: 'Generations of Irish women have left the country to buy contraceptives in Britain or Northern Ireland, or to undergo sterilisation and abortions, all prohibited at one time or another by law. A new generation is leaving for fertility treatment ...' (ibid.)

The assertion of a political mandate on regulation is framed with reference to the growth of a private NRT industry in Britain and the need to monitor clinics in an environment which is potentially exploitative of 'vulnerable couples desperate to have a child'. While the terms of the debate have changed, the normative framework remains the right of couples to have a genetic child of their own. In one newspaper article this is articulated as 'the basic human right to start a family' (*The Examiner*, 27 January 1996). This has become the central claim through which political claims about the allocation of scarce medical resources and the traditional hegemony of a Catholic ethos in the health domain are challenged.

The prospect of legal challenges to force a mandatory responsibility on the state is highlighted by the announcement of a High Court action to be brought on behalf of Gerardine Gleeson; co-operation between Cathy Murphy and the NFSIG in pursuing a case in Europe, and an alliance between Bourne Hall (a private fertility clinic in Britain) and the NFSIG to provide treatments in Ireland which are ethically banned at present. This alliance is explicitly framed as a mobilising strategy to force the pace of ethical and legislative change and it is contextualised by the personal narrative of fertility patients.

The issue of fertility treatment as a 'right' is being pursued through the alignment of legalistic and medical discourse frames. The fact that this is being brokered by an outside private clinic is, on the one hand, represented as an indictment on the Irish medical profession and, on the other hand, it serves to incorporate a market rationale on consumer choice and the need to protect clients in this unregulated market. The legal uncertainties which pertain in relation to prospective children born as a result of treatments involving donor gametes and surrogacy sought outside the Irish jurisdiction are negated in much the same way as the legal ambiguities of prospective AID children. This legal ambiguity is normalised by the common use of idioms such as 'genetic mother', 'host mother' and 'genetic father'. These encode a set of social relations which appear to have a symbolic order within a biological idiom, as well as a social order which can be adapted to the legal model of adoption.

Competing Discourse Frameworks

The coverage of the postmenopausal NRT story draws heavily on the textual references and discourse frames of the British media. The terms of the debate are mapped out by an expanding referential structure of argumentation which is part of the genealogy of NRT over its key discursive moments (as outlined by Van Dyck, 1995). These include: the well-being of prospective children; the encroachment on nature by science; the analytical or philosophical question as to how far science may be allowed to overstep normative boundaries ('We can now play God, but should we?') and the script of the narcissistic career woman, which is an extension of the 'designer baby' metaphor. While the anti-scripts of the 'unfit mother' and the 'Frankenstein' image of maverick scientists are part of the construction of a 'moral panic', competing interpretations are also voiced.

Within the 'expert' community those who oppose age restrictions in IVF programmes draw analogies with familial arrangements which are part of our cultural models. Take, for example, the following quotes: 'There are lots of grandmothers who cope with children. It is perfectly possible that the children can have a normal childhood' (*Irish Times*, 28 December 1993); 'Older women have always been providing such homes in Irish society. There is the child born to the teenage daughter tacked on to the family and reared by the mother; there is the tradition of one child of a large family given to the grandmother to rear. This has been going on in Irish society for hundreds of years' (*Irish Times*, 1 and 3 January 1994).

Contestation over the concept of 'normality' indicates the need for the inclusion of a wider and more diverse public voices. Discursive closure loses sight of the diversity of cultural values which shape how we perceive and interpret normative and analytical problems. In the debate on the social morality of older women parenting, the gender sub-text of the normative frame of reference is highlighted by oppositional voices. For example, the moral panic is counterbalanced by oppositional interpretations highlighting the nurturing role played by older women in the Irish familial context, which shows that cultural models have historically existed which attest to a non-threatening 'normal' parental context for rearing children. Narrative conventions for constructing subjective experience also create a dialogical context for an alternative collective social identity to emerge. For example, an *Irish Times* article, 'Too old to be a mother?' (1 and 3 January 1994), uses a self-narrative to construct the normality of older motherhood. Within the narrative the ideology of the modern nuclear family is inverted and challenged as the normative context which best suits the developmental needs of children.

Counter-myths

One of the more dominant intertextual references of the 1990s is the profile given to the archetypal female client of infertility treatments – the career woman in her mid-thirties. This profile highlights the mode of subjectification prevalent in medical discourse and it encodes new subjects as a counter-script to the definition of needs that arise from manifold experiences. The postmenopausal birth phenomenon has created a new category of Other which penetrates medical categories. Media coverage of Gail Sheehy's book *New Passages – Mapping Your Life Across Time* under the caption 'Fantasy of fertility forever' (*Irish Times*, 13 May 1996) is an 'anti-feminist' script which provides an explicit political sub-text for reading the administrative discourse of the narcissistic archetype. Sheehy's thesis is based on the 'illusion' of the 'feminist project' which has lulled women into a false sense of consciousness that mothering can eventually be planned for, but she argues that those women who had assumed control over their bodies – those who have been 'hoodwinked' into believing that their biology could be tailored to match their desires – are now left alone in their unproductive bodies.

This article is contextualised by a dramaturgical script set in a clinical consultation. Under the caption 'What do you mean, my eggs are too old?' the archetype is the wealthy successful career woman, who on the surface of her body has managed to reconstruct a youthful and healthy identity. The moral of the story is that her reconstructed bodily orientations cannot deceive the biological contingency of the procreative body: the wealthy career woman has to face the existentialist problem posed by the natural demise of her ovaries – she cannot transcend the nature of her own body. The script reduces the woman to an infantile state: her behaviour is portrayed as obsessive and irrational while the consultant in the reconstructed clinical setting represents the voice of rationality. The morality of science is framed in terms of a number of conditions set down in the script: science operates an internal logic which is assimilated to nature; this logic is operated by the rationality and morality of its brokers which protects scientific developments from becoming malleable objects of women's fantasies to overcome their own biological nature. Hence, science is not exploitative of women despite the demand of wealthy women to put themselves through medical procedures which are unlikely to yield successful results.

The proliferation of the narcissistic script can be viewed as a counter-factual discourse to the proliferation of the narrative convention of the experiential voice which has served as an impetus for public debate. It creates a new 'expert' discourse which attempts to reassert normative boundaries which can be easily assimilated into a political discourse based on the notion of a 'moral crisis' and at the same time can be assimilated and normalised by therapeutic and administrative modes of argumentation. While one may be tempted to argue that this counter-script provides a moral discourse against the scientific-marketisation rationale which has driven the international NRT debate, this would be overstating the point. Instead, this narrative may be seen as an extension of a common epidemiological script which serves to perpetuate a binary argumentation structure and a reductive political discourse. In the case of Ireland, where a context specific debate is beginning to emerge and where oppositional publics and counter-claims have yet to materialise fully in a political sense, this script may well be strategically deployed to negate potential fields of struggle and resistance.

CONCLUSION

While the NRT debate in Ireland cannot be mapped sequentially onto the trajectory of international discursive events, the international framework provides a contextual resource for the unfolding of the debate. On the one hand, the discursive legacy of key events and their meaning construction sustains self-referential systems in the construction of competing knowledge claims. The result of this, as Van Dyck (1995) shows, is that opposition discourses have been rendered invisible and the propensity is for public debate to become increasingly defined by a 'closed' and 'mechanistic' discourse constituted by a narrow range of 'eligible' voices – namely, those of the experts. Furthermore, the practice of political exclusion and closed referential structures 're-strict[s] the kind of questions that can be asked within current public debates' (Franklin, 1990, p. 226).

Expert cultures, however, are not homogeneous social fields of action, and while discursive and other resources can be more easily mobilised by institutional actors against less well-defined publics, cultural shifts occur which make possible new discourse alliances. While the earlier phase of debate in Ireland was structurally bound by the twin politics of abortion and familism, the cultural and social landscape has since shifted. Not only have new actors emerged to constitute a public debate, but life-world frameworks *vis-à-vis* changing narrative conventions in the con-struction of 'public opinion' are beginning to draw on wider cultural resources than traditional binary argumentation structures. For example, while abortion politics remains part of the interpretative framing of NRT, and its political lexicon central to how social actors locate them-selves within that discourse, new discourse frameworks are being cast. Since IVF has been normalised and indeed 'naturalised' within a rights idiom, the medical profession are less likely to be tied to moral frame-works which are increasingly being displaced by other rationales. The construction of infertility as a health status issue furthers the need for public debate requiring the state to act as a mandatory agent over competing rights claims.

While the state is reluctant to engage in a process which will foster further politicisation of the issues currently motivating the need for public debate, changing cultural practices within conventional political structures and within prefigurative politics renders a structured non-response less feasible than it was in the 1980s. The need for a public debate has a symbolic resonance within the politics of risk which has become a mobilising rhetoric within different discourse frames and by different discourse publics.

5 Biotechnological Communication and the Socio-Cultural Embeddedness of Economic Actors

Marion Dreyer

Companies and products are becoming more and more similar. Therefore, it is increasingly communicative competence which decides company success and failure. Communication is successful, when a company and its neighbourhood trust each other, when they seek a consensus between entrepreneurial activity and societal needs. Today, more than ever, companies have to deal with the communicative challenges of society. This implies taking responsibility, and admitting and correcting mistakes. Only in that way do companies have a chance to effectively shape public debate over benefits and risks, over rejection and acceptance. (Schönefeld, 1996, p. 72; trans. Dreyer)

INTRODUCTION

This chapter proposes the following hypothesis: in the genetic engineering controversy in Germany direct and indirect costs produced by public, political and judicial risk sensitivity have induced the chemical industry – as profit pioneer in modern biotechnologies and so far the industrial sector most reliant on these new technologies – to expand and diversify its society-related activities. The industrial sector reacts to a manifest and cost-producing technology-critical environment with an *expansion of public communication*, in general, and an *expansion of dialogic and direct (face-to-face) communication*, in particular. In the interactive dimension the increased preparedness for two-way communication is accompanied by a *heightened willingness to co-operate* with explicit or potential opponents of the technology at issue in relation to safety considerations for humanity and the environment. Two-way communication and co-operative behaviour aim at building credibility, trust and social acceptance for corporate/industry decisions by conveying a message of openness, transparency and a willingness to tackle and clear up controversial issues. The key sociopolitical message to be conveyed is industry's willingness to act responsibly. This is defined as an industrial policy to act in the public interest and make those decisions and follow those lines of action that are desirable in terms of the objectives and values of society. Public interest in this respect is not primarily defined in terms of 'material benefit', but essentially in terms of public safety, health and a clean environment.

The hypothesis holds further that industry's commitment to public dialogue and the public good operationalised in this manner is basically an attempt to (re)produce industry's *social legitimacy* as an institution which develops and uses technology in order to maintain optimal discretionary control over technology-political decision-making in conditions of *sociocultural change*. This change, the argument goes, consists in the breakdown of traditional societal consent to the effect that technical progress equals societal progress. Formulated in positive terms: it consists in the dissemination of a technology-critical perspective advanced by new social movements. What social and political actors demand is technical change that is *environmentally and socially compatible*. This demand has achieved considerable social and political influence in all those countries – Germany above all, but also the Netherlands, Denmark and Switzerland – in which powerful, enduring and recurrent technology-critical protest forms part of the process of institutionalising and professionalising the new social movements.

The hypothesis is developed in three sections. The first section sketches the analytical framework, the 'embeddedness perspective' of neo-institutionalist organisational sociology, in which the hypothesis is formulated. The second section presents the case study – how the chemical industry has responded publicly in the German genetic engineering controversy – as the hypothesis' empirical basis. The third section takes these empirical findings to formulate and discuss the claim that expansion and diversification of industrial public activity in this context reflect and are a response to new *sociocultural constraints*.

THE EMBEDDEDNESS PERSPECTIVE

The societal role of the traditional economic institution, the firm, and its body of representing interests, the industry trade association, has changed markedly in the last two decades. Today, government and the public both make high demands on the social responsibility of commercial organisations. These demands go well beyond the classic role of the provision of goods and services. The demand above all is that companies take into account the social, ecological and cultural ramifications of their decisions and thereby acknowledge highly valued social structures and values. This applies, in particular, to the course of development and the results of technical change. The social role of economic organisations, thus, is no longer viewed exclusively in terms of efficient economic behaviour, but increasingly in terms of the maintenance and observance of a *general welfare* and *public interest*, which goes beyond the satisfaction of material needs and interests. That is to say, commercial organisations since the 1970s have undergone a profound *process of politicisation*. This process has put the economic system under considerable strain – strain in terms of loss of credibility and trust, production delays because of public protest, misinvestment because of changing legal constraints, impending consumer withdrawal and demotivation of employees. In doing so, it has induced economic actors as well as social scientists increasingly to understand the industrial environment not just as economic and technological but as political, social and cultural as well. In theory and in practice a heightened awareness can currently be observed that economic organisations are far from detached from their not directly market-related external environments. This awareness is connected to the growing conviction that successful industrial production is increasingly dependent on non-market-specific factors.

One approach that provides the analytical tools to investigate the wider environmental embeddedness of economic actors and action is the '*neo-institutionalist*' *approach in organisational sociology*, as it has developed in the United States since the mid-1980s. With *embeddedness* as a central concept this approach underscores the important role of the wider societal environment – culture, social norms, conventions, society, the political system, organisational fields – in the definition of interests, in shaping structures, and in influencing the courses of action an organisation might develop.[1] This influence largely consists in providing the organisation with social legitimacy. Social legitimacy, according to the neo-institutionalist perspective, is a major determinant of the success of organisations together with efficient co-ordination and control of productive activities. While neo-institutionalism originally dealt almost exclusively with organisations in the public sector, more recent studies have expounded the way in which for-profit organisations are subjected and responsive to normative pressures. What they point out is that for-profit organisations have to be both economically efficient *and* socially legitimate in order to safeguard their growth and survival.

In this view, changing societal ideas of what constitutes responsible use of technology means that organisations need to document – in terms of structure and/or action – conformance to those ideas. They need also to document the *social appropriateness* of technology against these newly developed ideas and their underlying values. Reference to public communication, the following section will demonstrate, is a means to document social appropriateness; and the genetic engineering controversy provides evidence that societal support for economic operations and organisations cannot be taken for granted, but depends on how organisations respond to their broader context.

THE ECONOMY OF DIALOGUE

The chemical industry is the sector of the German economy that most utilizes genetic engineering techniques. It is simultaneously the industry most affected by social regulation and organised protest as well as the industry most actively involved in society-related activities in this technology domain. While the big three chemical enterprises (Bayer AG, BASF AG, Hoechst AG) pursue independent public activities, the bulk of this type of industrial activity is carried out collectively through the industry association the *Verband der Chemischen Industrie* (VCI). Due to the dominant position that the big three occupy in this association (Grant, Martinelli and Paterson, 1989, p. 76), action at company and collective levels have the same thrust. Industry's society-related activity discussed in the following pages thus represents company *and* association action.

In order to illustrate better the way in which the extent and nature of society-related activities are contextually determined, the case study also considers – peripherally – the preceding era when these activities barely existed. The period between industrial uptake of genetic engineering techniques – since about 1975 – and the latest regulatory policy measure contained in the amendment of the Genetic Engineering Act in 1993 can be divided into *four phases of industry (socio)-political activity*. These phases can be differentiated on the basis of the prominence of public communication and its argumentative and interactive nature. They shall be denoted as follows: (1) corporatist negotiations (1975–83), (2) organised parliamentary risk communication (1984–7), (3) public education (1988–90), and (4) public dialogue (1991–4). The empirical focus is on the latter two phases. It is only at the end of the 1980s that the chemical industry substantially 'went public'.

The first two phases represent a period of widespread public indifference. Industry relied almost exclusively on its strategic location in the system of political intermediation to safeguard its interests, and its (socio)-political activity took place in the arena of institutionalised politics. In the first phase, industry's (socio)-political activity was restricted to negotiations within a corporatist framework of policy formulation. Traditonally, big business, the state and organised labour have participated in corporatist negotiations, but on this occasion the academic research community joined them. Industry had a particularly good bargaining position for two reasons: (1) the field of modern biotechnology was a beneficiary of an interventionist industrial policy that was aimed at furthering high-technology development; and (2) there was hardly any public or oppositional parliamentary interest in the new technologies.

In the second phase, industry's negotiations 'behind closed doors' were complemented by negotiations in the half-public/half-private arena of the *Enquête-Kommission 'Chancen und Risiken der Gentechnologie'* (Inquiry Commission on Opportunities and Risks of Genetic Engineering). The parliamentary commission differed from the exclusive form of interest representation of corporatism by proportional representation of the parties and by the inclusion of a moral theologian, a social scientist and a lawyer. The Commission's early abandonment of fundamental, ethically and culturally grounded questions led with some exceptions to a proponderance of rather moderate political recommendations in the context of a still complacent public. Industrial (socio)-political activity in this phase continued to be restricted to activities in the executive and legislative dimensions of the political arena.

With the publication of the *Enquête-Kommission's* final report the recombinant DNA debate moved into the public arena and its focus shifted from economic benefits to potential ecological, social, health and moral risks. This shift was significantly advanced by increasing protest on the national as well as local level. Industry, eventually, 'went public' at a time when an extended and intensified political and public risk debate was increasingly incurring costs to industry in terms of damage to its public image, the loss of negotiating power vis-à-vis the state, and absorption of time and labour-power. The most direct costs, however, were those connected with procedural delays in production authorisation following on from decisions made by political and judicial authorities and associated bodies that were conscious of technological risk. In September 1988, the *Bundesrat* subjected the approval of production facilities using rDNA technologies to a public hearing procedure under the *Bundesimmissions-schutzgesetz* (Federal Emissions Control Law). Insufficient applications for licences by corporations, ignorance on the part of the authorities and objections of local citizens' groups obstructed the construction of genetechnological production facilities. Furthermore, in November 1989 the Administrative Court of Hesse approved rDNA production facilities, subject to the provision of an adequate legal basis, which to that point did not exist. The decision was substantiated by the assumption that at present risks posed to the human population and the environment from such plants could not be properly assessed and had the potential to jeopardise the right to life and freedom from injury of those in the vicinity. As other *Länder* fell in line behind this decision, it brought almost all genetic engineering projects in Germany to a halt.

In conditions of an increasingly turbulent and uncertain external environment including organised local protest, the absence of legal certainty, a developing network of non-governmental organisations critical of genetic engineering, an alerted lay public and mass media, and a quasi-production blockade – the chemical industry then expanded its political field of activity and entered the public arena.

The public communication that industry resorted to as a direct response to this 'situation of crisis' is distinguished by two strategies. The first is an education strategy which is designed to persuade by 'factual' information and is targeted at selected opinion leaders and the general public. The second strategy is confrontational, and is targeted at critics and organised protest.[2] Both strategies are primarily pursued at the firm level in the form of speeches given by management and key personnel in various public forums, discussion panels, the press, media interviews, congresses and via publications. While the education strategy relies on de-dramatisation, de-mystification and lowering the emotional tone of rDNA issues as its leading 'tactic', the confrontational strategy relies on stigmatisation of allegedly biased critics. Frequently applied epithets include: incompetent, incoherent, emotional, politically motivated, ideologically-driven. In short, industry's public performance is self-confident and assertive. It makes it quite plain that it is convinced that its viewpoint represents the only 'reasonable' position on the technical controversy. The following quotation from one of the managing directors of Bayer AG at a press conference on 'Gentechnik bei Bayer' captures well the situational 'analysis' and 'remedy' of the chemical industry at that time:

> Despite the promising opportunities which genetic engineering offers science and the economy, the acceptance of this technology by the

German public is alarmingly poor. The reasons for this are ignorance, misperceptions and fear. One of the main reasons is that the industrial genetic engineering which we use is conceptually mixed up with potential applications of human reproductive biology and other aspects of human medicine. Ideological fringe groups, fighting on principle against technological progress, contribute to a further distortion of public opinion. In order to improve necessary public acceptance of genetic engineering we consider it as indispensable to make this difficult and emotionally loaded subject transparent in an open dialogue. By that we want to factually explain which measures we take to ensure safety. This is an essential concern of our Genetic Engineering Press Forum today. (Büchel, 1989, p. 28; trans. Dreyer)

These strategies, however, have not proved very successful. The production blockade has been overcome; the climate of public opinion, however, remains critical and protest vociferous. As a consequence, the conviction grows that absence of public acceptance is less a *cognitive problem* – the result of missing, insufficient or misleading information – than a *political problem* – drawing from a lack of public confidence in the chemical industry's willingness to make those decisions and follow those lines of action that are desirable in terms of 'old' *and* 'new' values. It is the general *Wertewandel* – widely discussed and not only in the social sciences – which is increasingly regarded as the root of all evil. This change in values, it is assumed, has produced new expectations of industrial practice in general and industrial technology practice in particular, and in doing has destroyed the previous 'matter-of-course' public confidence in industry as a technology-developing and using institution. Public confidence, it is concluded, has to be systematically (re-)produced by demonstrating a general openness to changed societal expectations as well as a commitment to the value orientations that these expectations are based on – a commitment that includes admitting mistakes made in the past and demonstrating a willingness to adapt industrial practice in the future. The 'situation analysis' forming the basis of *confidence-building communication* as the new communicative strategy[3] – a situation analysis which concerns not just the rDNA controversy but the chemical industry's public exposure in general – is captured in the following statements by two prominent trade association representatives:

There are many signs that our fellow citizens within the meaning of the so-called value change want to lead a *new dialogue* on the chances and risks of industrial production. Essential preconditions for this are *transparency* and *capability to lead a dialogue*, as well as *willingness to compromise with all parties concerned.* (Weise, 1988, p. 66; emphasis in original; trans. Dreyer).

[Industrial] chemistry is expected to contribute decisively to the solution of the big problems of our time; to the provision of sufficient food for a rapidly growing humanity, for the combating of disease and suffering, as well as for the safeguarding of our affluence. Not least, [industrial] chemistry is asked to preserve an intact, natural environment which is worth living in. ... We have ... the uncontested scientific and technical competence to widely meet these expectations. ... However, it is not sufficient ... to be technically and economically successful. People want to be convinced of our readiness to learn from mistakes, of the responsibility that we take in our thinking and acting. To show this is the central concern of our policy of dialogue. (Strenger, 1991, quoted in Ueberhorst and de Man, 1992, p. 45; trans. Dreyer).

Confidence-building communication, in the mass media and in face-to-face communication, is designed to convey the message that industry values its *socio-political responsibility* as high as, or even higher than, its responsibility to the organisational goals of economic efficiency and profitability. A socio-politically responsible use of genetic engineering is defined as a technology policy that is committed to economic welfare as well as to a clean environment, to public safety and health – in terms of risk avoidance and improvement in quality – and to a public dialogue on this technology policy. Hence in the *Gentechnik-Leitlinien* (Genetic Engineering Guidelines) it claims: 'People's safety in the workplace and the protection of the environment are given priority in dealing with genetic engineering' (VCI, 1990, p. 5).

Industry-wide 'self-regulation' in the form of a self-imposed code of conduct is one means of documenting industry´s commitment to society's demand for socially and environmentally compatible technology; extensive risk research projects are another.[4] However, the principal means of documentation is the practice of 'going public', of voluntarily entering into a public dialogue in order to create a consensus on industry technology policy. This consists in demonstrating that industry has imposed *public accountability* on itself. The extensive use of the term *public dialogue* and the increased organisation of public communication as two-way communication, which is in contrast to educative information typically taking the form of unidirectional communication, is to demonstrate industry's willingness to justify its technology activity as against newly developed societal expectations and underlying value orientations.

Public dialogue as the overarching communication concept has existed since the mid-1990s. During this period the trade association, VCI, launched a new communication initiative entitled '*Chemie im Dialog*'. With this initiative 'dialogue' has advanced to one of three *Umwelt-Leitlinien* (environmental guidelines) of the chemical industry,[5] that is, from a minor tool of associative and corporate public communication to an overarching and integrating society-related industry strategy.[6] In 1991/2 genetic engineering had become one of the programme's key subjects – along with the environment and chlorine. From then on industrial communication on genetic engineering was framed in terms of a public dialogue on the risks and benefits of rDNA technology, and more extensively organised as direct communication with selected targets and the interested public. Corresponding to the VCI communication initiative dialogue takes place on two levels 'in order to demonstrate the chemical industry's willingness for dialogue on as broad a front as possible' (Mariacher, 1991, p. 30; trans. Dreyer).[7] This should take place on both the direct physical level and on the level of mass communication using advertising and television. The emphasis is on face-to-face communication – so-called *direkte Dialog- und Konsensinitiativen* (IGl, 1993, p. 3) – on immediacy and authenticity of encounters with industry's traditional and 'modern' stakeholders. VCI-organised events on the level of direct encounter include attention-grabbing *Großveranstaltungen* (major events). The central event here is the *Bundesweite Tag der offenen Tür* (the country-wide open day). Genetic engineering has been made a main subject of discussion. The first of these open days, held on 15 September 1990, was organised as the prelude to a new communication campaign. The VCI has also organised *Dialogveranstaltungen* (dialogue forums) for selected target groups. These target groups are those constituencies that shape public opinion – in particular, politicians and journalists – and groups that have influence and a high degree of credibility – in particular, church representatives, teachers, doctors, as well as environmental organisations. Furthermore, a series of regional discussions have been held. The targets of this 'dialogue on the spot' are the interested public and environmental consultants and the sector itself: industrial firms, producers, processors and purchasers (IGl, 1992, p. 24). Besides these events the VCI has organised so-called *Dialogpartner-Veranstaltungen* (dialogue partner forums) for the people who responded to invitations (see below). The groups were offered information on and discussion about genetic engineering and the opportunity to visit a genetic engineering plant (IGl 1991, pp. 3ff.).

On the level of mass media communication in print the campaign consists of advertisements employing the 'Chemie im Dialog' slogan and also, a response coupon and/or the slogan 'Ihre Meinung ist uns wichtig' (your opinion is important to us) which have been launched to demonstrate a willingness to enter into dialogue.[8] With the response-coupon readers were asked to enter into a personal dialogue on genetic engineering by letter, telephone or participation in so-called dialogue-actions. One ad was itself designed in the form of a dialogue: it was composed of industry's answers to questions voiced in public debate.

At the firm level the new society-related strategy is less often realised by large-scale, organised and publicised dialogue-events and *Dialoganzeigen*. Rather, it is executed by a generally greater willingness to speak out in public and a greater openness towards critics and the interested public. Hoechst AG has been the most innovative in this field. In summer 1993 the company established the 'Gesprächskreis Hoechster Nachbarn' (a conversational circle of Hoechst's neighbours, a local panel which is similar to the American community advisory panel; see Meister, 1996). The panel was not founded in the context of the rDNA debate, but as a consequence of a series of operative failures in standard practice (Kesselring, 1995). In this context it was founded and designed as a tool for crisis management. However, it came to be institutionalised as a 'local round-table' for discussion of company-related, topical safety, environmental and day-to-day questions (Barthe and Dreyer, 1995, pp. 87–92). As such it has also served as a forum to discuss issues related to genetic engineering.[9]

INDUSTRY GOES PUBLIC: 'STRATEGIC' ACTION UNDER SOCIO-CULTURAL CONSTRAINT

What this case study shows is that in conditions of high public exposure[10] the German chemical industry finally decided upon a strategy of wide-scale public communication in the form of a demonstration of the sector's commitment to enter into public dialogue[11] and to the public good, operationalised in the first place as public health and safety, and a clean environment. This strategy is not without risks as public communication elicits responses and opens a debate, the dynamic of which can never be fully controlled. The extent of the industry's public exposure which has induced it to take this risk is a country-specific trait that Germany shares with only a few other countries, such as the Netherlands, Switzerland and Denmark. These countries have been characterised since the 1970s by a strong movement which is critical of technology and an increasing dissemination of associated arguments and ideas in various institutions, manifested, for example, in a willingness to consider purely hypothetical risks to humanity and the environment in technology regulation or ethically-based regulation in the sphere of human genetics and reproduction technologies.

In the second half of the 1980s, German industry's public exposure included a policy of mobilising a range of political and social actors who are fundamentally opposed to the development and use of rDNA technology. Such a policy has been pursued by the Green Party as well as by a range of organisations affiliated to the ecology and women's movements and animal welfare groups. The Green/feminist/animal welfare technology opposition rejects genetic engineering not on the balance of potential benefits and risks but primarily on deontological, non-utilitarian grounds, attributing to all life-forms not only a right to exist but a dignity of their own which are made the principal criteria for human intercourse with nature. Rather than safety, political-ecological and social considerations, it emphasises as the basis of its opposition to technology the integrity and goodness of the 'natural' made into a moral norm for human behaviour. Industry's public exposure also comprises local protest by citizen action committees. Almost all of the big chemical companies involved in genetic engineering, or planning to do so, are confronted with such committees, who base their protest on the argument of the potentially disastrous effects that unintentional, i.e. accidental, or deliberate, e.g. for research purposes in the fields of chemistry and agriculture, release of genetically modified organisms could have for human health and ecological equilibrium. Public exposure also includes the high-profile attention that Parliament (in the form of the *Enquête-Kommission*) and the mass media direct at technology and its risks – the high mass media response must be understood as due in part to the increasing socio-political relevance of the issue of technical risk since the 1970s and partly due to the cultural-historical relevance of technology in view of the eugenic atrocities committed in the Third Reich – and finally, risk-sensitive administrative courts that rule in favour of concerned citizen action committees.

Morally grounded fundamentalist opposition, risk-sensitive local groups, political and legal institutions and a critical mass media response create a need for the chemical industry to justify and legitimate itself as public organisations and political authorities need to do. In Ulrich's (1977) terms, the chemical industry is turned into a *quasi-public institution*. It is confronted with critical environments which cannot be 'managed' by traditional means of control – power and money – but require a means of control that is positioned somewhere between 'moral suasion' and 'rational discourse' (Willke, 1983, p. 64). What is needed is a focused communicative effort, a special endeavour to convince, not merely the established players in the political game but a larger public that includes companies' neighbourhoods, opinion-makers such as journalists, teachers, church representatives and to some extent also vocal opponents and critics of technology; what needs to be utilised is the function of the public sphere as a *resonance context of legitimation.*[12]

Persuasiveness in its turn cannot be achieved by relying on scientific-technical information alone. With the breakdown of societal consent on the role of technology, information on technical matters and technical safety, without accompanying information on the social objectives and values on which safety and the very development and use of technology are based, is no longer sufficient. Persuasive power not only requires that these social objectives and values be articulated; it also requires a special commitment to the objectives and underlying value orientations of that idea whose diffusion originally shattered consensus. That is to say, it requires commitment to the new social movement-generated idea of *environmentally and socially compatible technical change* which respects human health and a clean environment as the central value orientations.

The chemical industry's commitment to this idea does not constitute a mechanistic reaction, so to speak. Instead, it is the result of a *learning process* in conditions of cultural change. The learning process consists of the realisation that a confrontational stance towards protest groups and a purely didactic approach to the interested/concerned public – both clinging to the traditional 'ideology of progress' – is not efficient. To put it into positive terms: the chemical industry experiences public commitment to newly developed convictions as a *socio-cultural constraint*. That does not mean that confidence-building communication does not *also* bank on 'materialist' value orientations. Rather, this new type of public communication represents of a new political identity, ideology and positions that incorporate different and sometimes conflicting societal objectives and values. Public dialogue is the cognitive and communicative effort of developing positive relations with a *heterogeneous* rather than homogeneous public interest.

While, on the one hand, the diffusion and stabilisation of the idea of environmentally and socially compatible technical change has plunged industrial organisations into a crisis of legitimation, at the same time, it provides the possibility of a way out: legitimating industrial technology activity and (re-)creating consent with society by means of framing it not only in terms of 'material welfare' but also in terms of environmental and social compatibility. The possibility of a way out is the possibility to use the latter idea as a *legitimating ideology.*

Is this to say that enhanced societal expectations can be created on the rhetorical level of 'symbolic politics'? To study organisational reaction to such demands in terms of political activity is to highlight organisational *agency*. It is to stress that economic organisations do not readily conform to this type of external dictate but act 'strategically' in terms of hiding from or negotiating it. Industries in the public eye recognise the framing of industrial practice and structure in terms of the idea of a socially and environmentally compatible technical change and negotiation within the framework of this idea as a political necessity. Political necessity, is deduced from the objective of keeping the influence of this idea on industrial practice and structure as minimal as possible. At the same time, communicative action based on this perception and aimed at achieving this objective produces some degree of self-restriction in the pursuit of organisational interests. In the final analysis, self-set normative bounds to a certain extent make restraint necessary. They (re-)produce normative expectations behind which industry's behaviour cannot fall without risking an even greater deficit of credibility. Publicly advanced self-commitments can be prosecuted by opponents at any time and with high public attention and moral pressure. The drawback of industry's political activity in the confidence-building mode consists in a further loss of credibility and legitimacy, when industrial practice does not fulfil what industrial communication promises, when it does not proceed on the basis of what it publicly propagates as the moral basis of entrepreneurial action. Sham dialogue and sham co-operation, therefore, even if it is designed as such, is only possible to a restricted extent.[13] In short, industry's 'strategic' communication produces a normative pressure for reflection as an action principle and with that an operatively restricted room for manoeuvre. This fact reveals the extent to which the dissolution of the *Fortschrittskonsens* (consensus on progress) operates as a constraint on industrial behaviour which extends beyond the impact of regulatory action alone. It means obligations on the level of both industrial practice *and* industrial communication.

6 Constructing Difference: Discourse Coalitions on Biotechnology in the Press

Patrick O'Mahony and Tracey Skillington

INTRODUCTION

Over the last 30 years, growing consciousness of the fact that regimes of nature are not eternal and unalterable has energised the idea of 'nature as victim' in human ethical and political consciousness. The depth of public reaction to biotechnological innovation shows society's increasing interest in this discourse. The once confident foundations of scientific-technical rationality are now questioned by a public whose increasing power to impose ethical standards on technical solutions has led to a gradual shifting of ethical rule-making capacities from the traditional institutionalised public narratives of government, business, education and science into the public sphere of modern society, where ideas and perspectives are subject to critical construction rather than inheritance (see Dreyer, this volume).[1] The employment of multiple and extended rationalities in assessing risk leads to opposing public constructions of biotechnological issues. These constructions are filtered through the dramaturgy of public display (Lowe and Morrison, 1984, pp. 75–90). The politics of biotechnology derives momentum from what Munch (1991, p. 62) describes as an inflation in the value of words, which provokes governments into making ethical and legal commitments that are often not adhered to, and compels industry to adjust its desire for innovative freedom to suit a more constraining symbolic environment (see Dreyer, this volume).

The analysis which follows explores this process within its wider political and cultural environment by representing the differing ethical perspectives and responses to recent bio-innovations by a wide range of social actors in the press between 1992 and 1996. The analysis follows the contention that scientific disputes are not resolved simply by reference to scientific 'facts', but by the adoption of rhetorical strategies that weave together ideological elements in a manner designed to shape public discourse and gain legitimation. The 'symbolic packaging' of evaluative and factual claims by like-minded 'discourse partners' has acquired some stability in the current debate on biotechnology. In this chapter, these positions are reconstructed by means of discourse analysis on a selected media bank. The majority of the 52 articles in the media bank are from Irish newspapers while the remainder are from British or American sources. Although this may seem to bias the findings of the research and limit its relevance primarily to the Irish context, the authors feel that the results are presented in such a way as to be of more general significance.

In the next section the theory of discourse-social connexity used in analysing the media bank is outlined. In the following section, the 'discourse coalitions' that carry distinctive points of view on biotechnological innovation are presented according to this theoretical framework. In the concluding section, an assessment is made of the respective influence of these coalitions on public discourse and institutional arrangements in the contemporary period.

DISCOURSE AS SOCIAL PRACTICE: A THEORY OF DISCOURSE-SOCIAL CONNEXITY

Fairclough's (1992) theory of discourse as social practice argues that discourse is not just about symbolic creativity since it also gives expression to ideological elements contained in what he calls 'plural rationalities' (see also Thompson, 1991). Through discourse, ideology is activated in its various roles, which include the explanatory, the evaluative, the orientative and the pragmatic. When utilised, these roles give an actor an ideological programme for social and political action. Therefore, how an actor presents an issue is not simply governed by immediate factors, but also by wider systems of meaning, or what McGee (1980) describes as 'ideographs', which are selectively and strategically called upon. Lemke (1985, 1988) describes ideology as a system of conjunctions and disjunctions that actors regularly employ to connect and disconnect social discourses and meaning, making practices in ways that function to maintain or undermine the given 'social semiotic system' (Thibault, 1991, p. 25).

While this research reveals how meaning is dislocated from stable discourses of cultural reproduction and dynamically relocated through meaning making discursive practices (Thibault, 1991, p. 44), innovations in the social construction of reality are never so transformative as to be culturally unrecognisable. There exists, prior to discursive practice, a preconstructed cultural setting within which any discourse is minimally anchored and through which it becomes comprehensible. This context, however, cannot wholly determine the contours of a public discourse. Hence, the effect of macro-structures – interpretively mobilised through institutionalised cultural codes – on discourse does not occur on the basis of clear lines of causality but rather on the basis of principles of connexity.[2] The concept of connexity may then be employed to understand the relationship between stable cultural meaning systems and creative episodes of discourse in a fashion that allows for their reciprocal determination.

This analysis investigates both intra-discursive connexity, where associative relations amongst meaning components internally organise a discourse, and extra-discursive connexity, where associative relations between text and context render it coherent through cultural proximity to broader ideological configurations. A four-tier model of discourse analysis is adapted from Fairclough's (1992) three-level model. The *preliminary level* will examine the perspective underlying each discourse coalition's internal discourse structure. The emphasis in this instance will be on local meaning relations and modes of symbolic deployment across actors in a coalition. What may be described as framing strategies, modes of representing characteristic points of view, organise the semiotic resources through which discourses are assembled and communicated. A *second level* examines how a process of intertextual construction between discourses leads to elaborations of existing discourses to give them greater scope and relevance. In this process, elements of different discourses are semantically combined by the selective foregrounding of certain meaning relations which reinforce a particular perspective. A *third level* examines how the discourse formations of a particular coalition are linked through a process of connexity to ideological resources in the wider social, cultural and political environment. Finally, a *fourth level* seeks to explain why certain coalitions' discourse formations enjoy a greater degree of institutional affirmation than others.

The focus of the research will be on discourse coalitions rather than discourse actors. Discourse coalitions are complex intersections of social meaning-making practices, comprised from the semantic registers of different social actors. The concept of coalition does not presuppose a strongly unified and fixed allegiance in which coalition members can be subsumed within one cultural framework over time. Instead, the shifting and contradictory nature of these alliances is indicative of a social dialectic encompassing a constant process of articulation, disarticulation, and rearticulation of discourse formations within a changing social and political environment. Solidarity between actors in a coalition may not exist outside of a specific context with the result that actors aligned on one issue may find themselves in conflict over others. There is, then, a dynamic patterning of the semantic relations in and through which systems of discourse are maintained or changed by members of various coalitions who operate within particular socio-historical conditions and ideological allegiances.

The analysis that follows proceeds from the coalition with the weakest principle of connexity with the social to that with the strongest. The first three levels of analysis are considered for each coalition. In the concluding section, we turn to the fourth level to explore how ideological currents prevalent here can be related to macro-social and political conditions.

DISCOURSE COALITIONS ON BIOTECHNOLOGY IN THE IRISH AND BRITISH PRESS IN THE 1990s

The Fundamentalist Critique of Biotechnology Coalition

In the case of the fundamentalist coalition, the first level of analysis reveals how the wording and framing strategies adopted employ a risk discourse and an anti-science discourse which together articulate a politics of fear.[3] Both these discourses, together with an anti-consumerist discourse, collectively communicate a cultural theme of destruction, captured in allusions to the 'sinful' nature of the manufacture of 'à la carte humans' and other bio-innovations that reduce 'life to its marketability'.[4] Thematic connectors facilitate the communication of a cultural theme of exploitation, and in this instance, include a strong reliance on verbs such as 'terrify', 'frighten' or 'mutate', nouns like 'pests', 'horror', 'threat', 'unnatural', and adjectives, including 'vulnerable', 'deadly' and 'sinister'.[5] Intra-textual connexity is constructed from the associative relations between elements of meaning that unify the coalition's negative typing. For example, to complement the cultural themes just outlined, analogies with 'Hitler' or Huxley's 'Brave New World', as well as tendencies towards 'madness' are projected as co-referentials with the modern genetic scientist and her research.[6]

The fundamentalist coalition's framing of humanity's moral responsibility toward nature is very distinctive. It links this issue to basic cognitive structures of modernity which embody a nature-dominating code. As against this, an ethic of care towards nature is promoted through identifying general principles of equity that should apply to humanity, animals and the rest of nature. This coalition's forceful defence of animals' rights criticises instrumentalist approaches in which an animal's worth is measured in terms of its utility to humanity.[7] The discourse of moral responsibility encompasses a secondary frame of just distribution, one which has progressed beyond a materialist concern with fair distribution to considering distributive justice for non-human nature. The adoption of the animal rights discourse and complementary cultural themes is constrasted to themes of opposing discourses such as economic expansion and faith in science. The opposed others are those scientific-industrial actors responsible for the morally unacceptable technologisation of food, a technologisation propelled by the profit principle perceived to be driving forward biotechnological initiatives at an irresponsibly quick pace.

The second level of analysis of this coalition reveals intertextual borrowing that results in modifications to the existing discourse. A particularly prominent example is the fundamentalist coalition's creative adoption of a non-materialist, secondary frame of distributive justice. This secondary frame is an outgrowth of the master frame of materialist distributive justice that primarily became institutionalised with the founding of the welfare state. An extension of its ideational elements and an injection of the counter-cultural theme of harmony with nature extends the notion of the common good. The fundamentalist coalition embeds this frame in an altered discourse of rights which extends the concept of justice to the animal community. Its sense of injustice with regard to biotechnology's treatment of animals is culturally derived from the transferability to them of human entitlements. The new theme provides a normative standard for criticising the exploitation of animals through bio-innovation.

The example above illustrates how textual production can be a creative process as much as it can be situated within existing cultural patterns. In terms of 'accredited sources' (Hall et al., 1978), journalists in this particular coalition predominantly quote influential environmental protest actors from CIWF, Greenpeace or the Green Party. Quoting such sources involves a recontextualisation in which concern is less with the representation of a quoted actor's discourse than with meta-communicative information about the dialectic of insider (the journalist) and outsider (the quoted official) relations and the joint context they enact. Quoted points of view function through imposing constraints on the production of alternative meaning.[8] Collectively, these give rise to a relatively coherent 'voice(s)' that distinguishes this coalition from that of others. Thibault (1991) borrows the concept of voice from the writings of Bakhtin (1973, 1981) and Volosinov (1973) to show how specific meaning constructions generally correspond to the position of the actors that constitute a specific coalition and can therefore be institutionally located. The 'dialogic' relations amongst the fundamentalist coalition's textual voices indexes a variety of ideological positions at the same time. This discourse alliance's positioning in the intersection of animal rights, risk and anti-consumerist discourses, a secondary distributive justice frame, their typical co-patternings, and the interests served by these symbolic strategies, determine how this coalition constructs meaning.

In terms of relating to a wider set of cultural beliefs. At the third level of analysis, this coalition endorses a conception of the common good as 'stewardship'. From this perspective, humanity is not seen as privileged in relation to other life forms and contemporary society is portrayed as having the role of carefully managing natural resources for present and future generations (Sessions and Naess, 1991). Without being beyond the reach of politics, the dividing line between the cultural and political aspects of this alliance's discourse formations are the least politicised, in the sense of being born out of societally confirmed political cleavages. Capitalist society's capacity for the semantic register of an alternative cultural reasoning is thought to be outweighed by its commitment to an economist rationality which depends upon the subservience of nature to humanity's needs.

New Left Libertarian Critique of Biotechnology Coalition

A first level of analysis of this coalition's perspective shows a fusion of the socialist frame of international equity with that of environmental protection. Equity is linked to the need for tight, ethically guided government regulation and to the avoidance of economic exploitation of indigenous farmers and those in the Third World.[9] The coalition calls for 'intervention' to protect the autonomy of 'farmers' who would otherwise become 'major victims' of biotechnology patenting.[10] Equity-oriented intervention is projected as a counter-theme to that of corporatist-styled decision-making with its premises of elite pluralism, consensus and, ultimately, benefits.[11] Similar to the fundamentalist coalition, this alliance espouses a discourse of environmental protection. This discourse is designed to counter the biotechnology as solution coalition (see below) and the latter's attempt to equate 'sustainable agriculture' and food biotechnology, by arguing that 'genetic engineering' involves the exploitation of nature rather than co-operation with it.[12] A frame of distributive justice is used here as in the fundamentalist case, but its formulation is a materialist one – in this instance centred on the concept of equitable international trade relations. Although critical of the manner in which recent biotechnology innovations are being steered, this coalition is not opposed to science *per se*. There is a general acknowledgement that science is of merit to humanity in its ability to save lives and alleviate suffering, but this is tempered with the belief that its application needs high regulatory intervention as scientists are 'notoriously bad at disclosing the truth'.[13] The mixed combination of symbolic strategies employed by this coalition links moral discourses about the environment to socialist ideas of distributive justice within the politicised framework of defining nature's limits and a global conception of the common good.

In terms of the second level of analysis examining intertextual transformations between discourses, this coalition's materialist framing of distributive justice has a close affinity to the primary frame of just distribution. Concern, in this instance, is with international equity in terms of trade with poorer countries. An intertextual connexity is established between this coalition's reactive frames of meaning and its formulation of a meta discourse of democracy that emphasises 'true public debate'.[14] Goffredo del Bino, chief of biotechnology regulation in DGXI, is quoted in his reference to 'transparency in biotechnology regulation' as a prerequisite for 'taking it in a democratic direction'.[15] Quoting this official generally attempts to create an impression of a synchronisation between the reporter and the reported and is a form of 'manifest intertextuality' (Simons, 1995) where the official's text is responded to, reinterpreted or reinforced in an effort to increase the perceived cogency of the argument being presented.

The coalition's critical perspective on the need for greater transparency and dialogue with the public at large shows how a reading of a more 'discursive' democracy, part of a revitalisation of debate on democracy on the Left, is used to interpret and evaluate the new biotechnology era in a high-tech capitalist environment. Enhanced discursivity will make even greater complexity possible by preserving innovation 'goods' and eliminating 'bads'. A substantial theme linked to enhanced democratic discourse is that democratised criteria of distributive justice must be used in addressing international equity and other collective good problems brought into relief by biotechnological innovation.

A third level of analysis, addressing discourse-social connexity, shows how, unlike the fundamentalists' discourse coalition, the socio-structural basis of this coalition has emerged from long-run change. Political opposition, originally inspired by a socialist perspective and associated 'repertoires of contention', facilitated the coupling of processes of technical innovation with those of politicisation, mobilisation and ensuing democratisation. The sediments of past political mobilisations continue to shape new protest cycles (Tarrow, 1989) through strategic actors' symbolic framing devices which remain as ideological as those of their predecessors in the sense of conceptions of the right, the real, and the meaningful (Eder, 1992, p. 20). This coalition's use of a distributive justice frame is inspired by a deeply rooted tension between societal commitments to effective performance and to equality, which together structure modern society. Such tension is expressed through fundamental antinomies, including trust versus mistrust in capitalist innovation processes, and elitism versus egalitarianism as ethical constructs. The capitalist commitment to effective performance produces inevitable inequalities, particularly in this perception with respect to the distribution of knowledge and risk as well as resources. The principle of equality is intermittently invoked here to challenge the legitimacy of such inequality, but is situated within a broader context of ambivalence about not losing the economic and, to a lesser extent, the social benefits of biotechnology.

Counter-scientific Expertise Coalition

The existence of a counter-scientific expertise coalition indicates both a divided scientific community and contradictory factual opinions on the benefits and risks of biotechnology. A qualified belief in the potential for empirically grounded objectivity in science remains significant, but this can no longer be assumed to be an undisputed axiom guiding all science in all situations. Hence, as the first level of analysis of this coalition reveals, the cultural theme of faith in science is undermined through, paradoxically, science's own production of counter scientific 'facts' on the effects and repercussions of biotechnology, a process that leads to the increasing salience of public discourse to address controversy. This coalition embraces the precautionary principle, i.e. that no innovation that might result in potential harm to humanity or the environment should proceed seriously on the basis of research findings that demonstrate how such harm could ensue from certain biotechnological innovations. It thus recommends caution when the release of genetically modified organisms is under consideration and argues in favour of scientific assessment according to the worst-case scenario in terms of impact on the environment.[16] It regards the hallmark of disinterestedness as inappropriate to the social organisation of science. The fusion of scientific arguments and moral injunctions is represented in the loss of a clear distinction between 'rational' propositions about the threats posed by biotechnology on the one hand, and normative propositions expressing doubt about its general implications on the other. The lack of distinctiveness between rational and normative stances is reflected in this coalition's articulation of anxiety and scepticism about such contemporary innovations as 'tissue engineering', involving the fabrication of human body parts using human cells grown on animals, or 'sheep cloning' and the ethical dilemmas they give rise to.[17] Ambiguity amongst experts about scientific 'progress' inevitably exposes science to the dynamics of ethical discourses.

A second level of analysis demonstrates how quotes from leading medical and scientific professional bodies are utilised to give greater coherence and cogency to the thematic structures of texts. Professor David Hopkinson, Director of the Human Biochemical Genetics Unit at University College, London, is quoted in his disapproval of the practice of sheep cloning for the sake of putting 'an extra fiver in the farmer's pocket', as is the Rev. Dr John Polking, President of Queens' College, Cambridge, and Chair of the Church of England's Science, Medicine and Technology Committee. Inconclusiveness is constituted through a process of intertextuality that combines a discourse of scientific objectivity with one of environmental protection in an ambiguous way by introducing a theme of biotechnology as problem. The second level of analysis shows the coexistence of a variety of discourse types and cultural themes which convey varied voicings, including that of a qualified faith in science, yet expressing the need for caution when evaluating its findings.

The fact that the traditional distinction between knowing and valuing within scientific epistemology is undermined by scientists makes the critique of science all the more damaging. The third level of analysis highlights how technical-utilitarian rules that previously determined exchange between nature and society have become less clear-cut and subject to scrutiny. The self-evident assumption that nature is an infinite reservoir on which the effects of biotechnology research will always be negligible is questioned. This is evident in this coalition's sharp criticisms of certain scientists in the field on the grounds, for example, of underestimating the effects of the release into the environment of unwanted target organisms from genetically modified plants. Similarly, critique is directed at regulation which fails to consider the needs of society rather than merely those of research. Support for this discourse coalition is growing as the politics of uncertainty and risk increasingly has an impact upon the regulatory climate.

Biotechnology as Solution Discourse Coalition

The first level of analysis of biotechnology as solution coalition shows how the extensive use of nominalised verbs and the passive tense presents opposition to biotechnology as lacking in authority. For example, a typical statement is that non-scientists and concerned citizen groups do 'not have a complete understanding of the context in which the technology will be developed and ultimately used'.[18] The coalition presents itself as devoid of irrational motivations, as having no impulse towards public self-assertion and political gain. This presentation of the self is portrayed as the antithesis of opponents to biotechnology who 'exaggerate' the public's fears about a science that is 'mundane and aimed at developing health-enhancing products'.[19] On occasions, criticisms of opponents surface with 'anti-food additive outfits' being conveyed as irrational 'car bombers' and 'terrorists'.[20] This type of referencing system is very similar to that observed by Altimore (1980, p. 25), who studied articles on the recombinant DNA controversy which appeared in the *New York Times* and the *Washington Post* between 1974 and 1980. He discovered that reporters predominantly quoted scientists in favour of the research, who would stand on their authority to dismiss rDNA adversaries as 'zany'. Opponents of biotechnology are portrayed as having an 'irrational' reverence for untampered nature. Critique is driven by the underlying premise of the truth-claims of empirical science, a perspective that is, in uncertain cases, in reality not based on hypothetical but normative argument (Habermas, 1974, p. 269).

It is in this coalition's symbolic negation of its opponents that material or political intent becomes apparent. Its disconfirmation of the other – usually environmental or animal rights activists with a point of view on biotechnology – is not concerned with establishing the falsity of the non-scientist's opinion, but with the negation of the reality of the non-scientist as a source of any valid opinion on biotechnology. By contrast, scientists within this coalition claim to be working on behalf of the needs and interests of the people. The adoption of the rhetoric of 'the people' and similar ideographs that embody and yet disguise a broad constellation of politically functioning interests and values is called upon in an effort to secure legitimacy (McGee, 1980).

A second level of analysis shows how frequent use is made by members of this coalition of the master frame of sustainable development, an important component of ecological modernisation or ecomodernist discourse (Hajer, 1995). The latter represents pollution as something that can be dealt with through the use of 'environmentally sound technologies'. Biotechnology is seen as one such technology in making possible 'a natural, environmentally friendly approach to food production'.[21] An ecomodernist discourse is adopted to promote the concept of 'biocontrol', that is, the use of bacteria on growing crops to counter a 'dangerous over-reliance on chemicals of all kinds to subdue the multitudes of pests which attack food' and hence 'suppress the development of harmful diseases'.[22]

The formulation of the frame of sustainable development is structured from the perspective of an economistic rationality, which is taken to be the norm. The sustainable development frame, in this instance, is set within a modernist meta-cultural discourse that presents the unification of the capitalist concern with profit and expansion with 'environmental care'.[23] The concept of register allows one to relate a coalition's configurations of semantic selections or textual 'voicings' on sustainable development to higher-order intertextual practices that reveal a symbolic struggle to define or redefine the copatternings of social meaning making practices (Skillington, 1997 p. 15). It is at this level that it is possible to detect the types of political manipulation the frame of sustainable development is subject to in order to suit particularistic interests. In the case of this coalition, sustainable development is made consistent with techno-economic innovation. This has the double function of opening up the environment as an economic opportunity and playing down anxieties about intervening in nature.

By 1995, this coalition had also, through a similar process of tactical innovation, adopted the frame of material distributive justice from the left-libertarian oppositional coalition and used an altered version of it to promote agricultural biotechnology that will supposedly allow farmers in the developing world to license 'cheaply' genetically altered seeds and plants.[24] In another variation, Professor Avigdor Cahaner, an Israeli geneticist, argues that advances such as his 'featherless chicken' 'are essential to feed the hungry, growing populations of the Third World countries'.[25] This coalition's adoption of this frame reflects careful judgement on their part as to what symbolic devices resonate meaningfully with the public, even if it fails to mention that considerations of profitability may limit the transfer of these technologies to the developing world.

It is noticeable that this discourse coalition does not depart from the dominant, nature-exploiting cultural code of modernity. It embraces a technicised notion of progress and a cultural theme of faith in science which emphasises the merits of the £100 billion a year biotechnology industry which promises rapid economic expansion.[26] Its optimism is captured in descriptive categories used to describe biotechnological innovation – for example, the 'impressive' contribution it has made to the 'dazzling American success story in science, medicine, agriculture and global commerce'.[27] It speaks enthusiastically of 'supercows' expected to produce 2000 gallons of milk a year, long-life 'flavor savor tomatoes', the officially approved science of xenotransplantation (transplanting organs from animals to humans), and even of cellular xenografts (the transplant of animal cells to humans).[28] Organs-on-demand is presented as the future utopian scenario: 'an unending source of hearts, lungs, kidneys and livers for human recipients'.[29] A strong commitment to 'inventions' of these kinds is coupled with a pervasive anthropocentric reasoning. As one scientist puts it: 'The whole point of using animals', he says, 'is for the benefit of mankind.'[30] The meta-level of symbolic power to define the contours of discourse on biotechnology in a nature-exploiting manner is not so much present as displaced to the implicit agenda-setting tactics of contextualisation. As Fairclough (1992) points out, discourse formations do not so much complement wider systems of cultural or political belief as respond to them in a creative manner.

A third level of analysis indicates the strength of the discourse-social connexity of this coalition. Strong institutional support exists for its faith in the age of 'biocontrol', which is seen as an incremental addition or 'one step further' to the current technological paradigm rather than a structural transformation.[31] It is assumed that biotechnology can satisfy material needs on an unprecedented scale and produce social cohesion internationally. Scientists know that seeking to create a pro-biotechnological consciousness is not going to secure public loyalty, but actual or likely gratification in the form of biotechnological consumables will.[32] Hence, the ideology of the market with its jargon of the 'new and improved' is called upon in order to justify these nature-altering developments. This facilitates the movement of this coalition's discourse from private interest to public good. Bio-technocratic ideology is frequently wrongly interpreted as merely as instrumental rationality when it has, in fact, real material interests and political ambitions. As Gouldner (1976, p. 261) explains: 'The technocratic consciousness and legitimating of modern society is fundamentally an emendation and a fulfilment of the grammar of ideology, rather than the end of ideology.' The emergence of protest, however, has disrupted this project, making instrumental reasoning ethically questionable and visible, thereby calling for, amongst the biotechnology community, a value rationality, that is, a moral commitment to some set of cultural and political ideals that can be argued for or against by a concerned public.

EXAMINING POLITICAL AND CULTURAL OPPORTUNITY STRUCTURES FOR THE FOUR DISCOURSE COALITIONS

The third level of discourse analysis of the four coalitions on biotechnology analysed above demonstrates how the discourse formations of each coalition are aligned with wider systems of belief by way of a principle of connexity. This fourth level of analysis, building on the previous three, further tries to examine the extent of institutional significance of each of these coalitions. Such coalitions could not exist without at least some measure of support in public culture. Potentials for public support measure its political and cultural opportunity structures.[33] Opportunity structures, in our usage, define the degree of openness or closure a particular institutional order exhibits towards a particular coalition.

The fundamentalist discourse at present enjoys considerably more cultural power than political influence. Its alternative political reasoning is codified according to a cultural orientation and a 'lifestyle politics' that attempts to open up symbolic space for a new conceptualisation of the common good. The associative relations prevailing between the semantics of this coalition's discourse formations (e.g. a belief in the rights of animals to live out their existence with as little intrusion from humankind as possible) and institutionalised cultural semantics (e.g. anthropocentric reasoning) are weak. In addition, its strongly moralised messages tend not to be carefully politically packaged. Hence, of the four coalitions identified here, the fundamentalist critique of biotechnology has the weakest degree of institutional effect, notwithstanding its significant public resonance. It is possible that if some bio-ethical controversies acquire a more stable and institutionalised form, then some fundamentalist positions may move from current stances to more influential, compromise-oriented positions. That something like this is already happening in Germany is clear from Dreyer's essay in this volume.

The new left-libertarian coalition is partly a response to the new, mercantilist climate in international trade where the advantage of nations is given priority, partly owing to a battle for supremacy in the new, science-based industries. It is also positively associated with the idea of distributive justice inherent in the global ethics of some conceptualisations of sustainable development. But this conjunction, fully elaborated in a new ecological politics of development, points in a direction too radical for the core of the coalition. Furthermore, while ideas of global distributive justice allied to environmental ethics have significant cultural resonance in western public culture, the ideology of materialist distributive justice is a fragmented one since separate criteria apply to 'third world' as opposed to domestic issues. Opportunity structures for this coalition tend on the whole to be more cultural than political.

The counter-scientific expertise coalition has increasing institutional significance in that ethics committees such as Warnock in the UK (see Barnes, this volume) increasingly decide on legitimate practices. While it currently has greater cultural than institutional power, it could rapidly gain in institutional significance, requiring only a catastrophe or major controversy to gain political leverage.

Notwithstanding, the other science-based coalition, the biotechnology-as-solution coalition still gains advantage from the current tendency, under the influence of liberalisation, to equate scientific-technical and economic good with the collective good. Given its present level of political and social affirmation, this coalition can symbolically absorb other competing arguments by deflecting them in a collective economic interest direction. This strategy is oriented towards the containment of controversy, while decisions affecting innovation, especially in less controversial biotechnological areas, take place without the public being able to scrutinise them. The coalition, therefore, relies on acquiring a covert institutional leverage that tries to ignore value conflict. On the whole, its political opportunities, while greater than its contested cultural status, is still circumscribed by the absence of legitimacy.

The overall impression conveyed by the account of the opportunity structures facing the four coalitions and by the analysis as a whole is that there is significant cultural distance between the different coalitions. Given the significant cultural power of each of the coalitions, consensus or how to deal with controversial biotechnological innovations appears far away. The affinity between the biotechnology-as-solution coalition and institutionalised economic objectives – and also social ones (see Skillington, this volume) – gives this coalition sigificant leverage over material resources. But its cultural power is far less than this and constantly likely to be undermined anew as further scientific frontiers are crossed with immense practical and moral implications.

Part III

The Dynamics of Institutationalisation: Juridification and Regulation of Biotechnology

7 Public Representation and the Legal Regulation of Assisted Conception in Britain

John Murphy

INTRODUCTION

Biotechnology makes strides ever forwards. Not always, however, are such scientific advances necessarily to be welcomed or even, sometimes, tolerated.[1] This chapter focuses on one such development – the revolution in human reproductive technology. When Louise Brown, the world's first 'test-tube' baby, was born in the late 1970s the very fact of her birth captured the headlines. Today, so commonplace are such techniques – most notably in vitro fertilisation[2] – that such occurrences have become positively mundane. Indeed, it takes a case such as that of Mandy Allwood, a woman pregnant with octuplets following IVF treatment, for assisted conception again to dominate the headlines. And yet, whilst laboratory-based births have become if not common then unremarkable, it would be false to assume that society has accommodated the new reproductive age in its stride. Indeed, it is the thesis here that any such reproductive revolution will produce a series of challenges for the law; that as medical science increasingly enables us to harness nature and 'play God', so must the law, in turn, attempt appropriately to harness medical science.

The first legal challenge is to devise an appropriate framework within which the acceptable limits of such scientific enterprises can be demarcated. To do this, it is necessary to recognise the crucial tension between morality (even, perhaps, ethics) on the one hand, and scientific potentiality on the other.[3] *Ex necessitate*, this tension is characterised by two central questions: namely, 'is it scientifically possible to achieve X?' and, assuming it is possible, 'ought we to permit X to be done?'. Secondly, assuming that assisted conception is permitted, the law must also provide at least a minimal series of rules. These rules, it will be argued, are required not so much *to curb* the practice of assisting the infertile to have children, but

117

rather to deal with the inevitable legal issues that arise upon, and are inextricably linked to, an assisted birth.

The scheme of this chapter, then, is to consider first the overarching question of whether the reproductive potentialities generated in research laboratories ought to be confined to the ranks of the merely feasible,[4] or whether they should become simply another aspect of everyday health care provision. Secondly, we explore the way in which the law ought to respond to the potential distortion of familial relationships, and the novel connotations of parenthood,[5] to which biotechnological reproduction inevitably gives rise. Thirdly, we address the confidentiality issues that stem from the fact that those who donate gametes[6] to enable others to have children normally wish to remain ignorant of, and anonymous to, those who ultimately receive those gametes and the children born in consequence of their use. Fourthly, we consider, on the assumption that something might go wrong with assisted conception, whether those clinical errors have the potential and moral basis to found an action for compensatory damages. Fifthly, we examine the thorny question of what can and should be permissibly done with any spare embryos generated by a cycle of IVF. Finally, we consider the appropriate regulatory framework to govern biotechnologically assisted conception and embryology. In relation to each issue, the English legal position will be used, largely for reasons of convenience, as a platform for discussion. That discussion will proceed on the basis of critical evaluation of the range of possible legal responses to each of the problems outlined and the appropriateness (or otherwise) of the English law.

THE RIGHTS AND WRONGS OF ASSISTED CONCEPTION

The question of how much liberty there should be to make use of the reproductive technologies that have emerged is not just of theoretical significance, but also, increasingly, of practical importance. According to one study (Douglas, 1991), between August and December 1991, over 6000 women sought IVF treatment in the UK alone.[7] Whatever the corresponding figures for 1998 will disclose, there can be no doubt that such treatment is sought by a very significant number of people.[8] That the number should be so great is in large measure explained by reference to the fact that demand for such services extends beyond the ranks of the infertile population.[9] Others, for example, the 'merely childless', might equally be unable to conceive without assistance. As regards the childless, however, the inability is attributable either to a psycho-sexual factor[10] or to some other biological impediment.[11] Similarly, those driven by eugenic concerns – for example, the desire to eliminate the prospect of giving birth to a disabled child or one of the 'wrong gender' – might also wish to make use of assisted conception. But we begin to run ahead of ourselves. For now, we are concerned only with the fundamental question of whether reproductive technologies[12] ought legally to be available in the first place.

There is indubitably a strong argument that the UK's hand has already been forced in the direction of the non-proscription of scientifically assisted conception. This argument is founded upon Article 12 of the European Convention on Human Rights and Fundamental Freedoms, which provides a *legal right* to all citizens of marriageable age to marry and *found a family*.[13] It could be argued, then, that any attempt to prevent someone from founding a family, by whatever means, amounts to an abrogation of Article 12. A counter-argument, however, might be that the right to found a family must be read *conjunctively* with the right to marry. If this interpretation is correct, then, so far as it relates to the right to found a family, Article 12 must be read as conferring no more than a right to raise *legitimate* children who were conceived 'naturally'.[14] If, however, the two limbs of Article 12 are construed *disjunctively*, there must exist an independent right to have children *by whatever means*. In this case, no restriction on access to reproductive technologies could be justified.[15]

There are two further grounds upon which it might be argued that assisted conception should remain unencumbered by any restrictive regulatory framework. First, there is no clear basis for the state claiming a right to interfere with the right to reproduce.[16] So called 'test-tube' babies attract an interest amongst the public, but their creation fails to engender any legitimate *public interest* sufficient to outweigh the private interests at stake.[17] Secondly, an argument based on property rights might be deployed to reject legislative constraints on reproductive freedom. Essentially, the argument runs that, in the case of IVF, as the gametes are taken from the childless couple, the embryo thereby formed is thus also logically theirs, and so reproduction by use of assisted conceptive techniques is merely one expression of their property rights.[18]

If this argument is correct, and biotechnological conception is simply a property issue, then it is pertinent to enquire whether the state has *any* legitimate basis on which to impose restrictions on the freedom to use reproductive technologies. Two suggest themselves. First, unregulated assisted conception inevitably raises the spectre of human cloning and the creation of creatures that are part-human and part-animal. English law absolutely prohibits human cloning.[19] It also largely outlaws placing a human embryo in any animal.[20] A limited inroad exists in relation to mixing human and animal gametes (so long as this is done purely for the purposes of research and on the condition that anything thereby produced is destroyed not later than the two-cell stage).[21] Secondly, there is a strong claim that the imposition of regulation can help ensure that assisted conception is performed professionally and with minimal risk of medical mishap: biotechnological conception involves delicate surgical procedures and requires suitably qualified professionals.[22]

Given that the prevailing European human rights principles can be argued to support a right to found a family – perhaps even using new technologies – it is unsurprising to find that IVF is permissible in the UK and in some health authorities is available on the National Health Service.[23] Once it is accepted that reproduction by means of artificial, biotechnological techniques is acceptable, it becomes necessary to explore those matters which, as a corollary, require regulation. In the remainder of this essay I examine each of these problems, outlined earlier, that demand such a response.

STATUS PROBLEMS

Historically, there was a necessary link between gestational and genetic motherhood, but following the reproductive revolution this is no longer the case. As regards the question of maternity, there will normally be at least two women – the genetic mother and the social mother – who will have a plausible claim to be treated for legal purposes as the mother of an IVF child.[24] Similar problems exist with respect to paternity. There are two men upon whom the badge of legal fatherhood could conceivably be pinned: the man whose sperm was used to create the in vitro embryo, and the child's social father. For the first time, then, especially complicated disputes over parenthood have become possible. The factual (as opposed to legal) multiplicity of 'parents' has given rise to a challenge for lawyers which is both novel and unavoidable. The novelty of the challenge, is entirely a by-product of the scientific potential that has emerged in recent years. The unavoidability of the challenge, however, is in part to be ascribed to human nature,[25] and in part to the pervasiveness of legislation that imposes upon parents the obligation to maintain and support their children.[26] As regards assisted conception, where two men could claim to be the child's father, it is far from obvious upon which of them this status ought to be conferred. If the law took the stance that the genetic link was the crucial factor, then, in Britain at least, sperm donors, in repayment of their altruism, would find themselves potentially liable to make child support payments.[27] This would not only deter men from donating their sperm but, more fundamentally, deny to those who fulfil the role of social parent all the 'natural' rights in relation to children that accrue by virtue of legal parenthood.[28]

Against this background, it is clear that once IVF and other such techniques become legally permissible forms of reproduction, the law must inevitably resolve the question of how to ascribe parenthood, and on what basis. The English response has been, largely, to confer this status on the *social* parents and treat them as the *legal* parents to the exclusion of all other persons,[29] so that gamete donors will escape any potential liabilities towards any resulting children. As observed earlier, one of the reasons for taking this stance is to ensure that sperm and ova donation are not discouraged. In addition, however, the English approach enables the social parents – i.e. those with day-to-day care of the child – to be in a position legally to take important decisions affecting the child's life (such as whether to grant consent to medical procedures – be they therapeutic or not).[30] In short, the rationale for treating the social parents as the legal parents is pragmatism: there would be very little point, or benefit for the child, in investing unknown genetic parents with the powers and duties which characterise legal parenthood.

CONFIDENTIALITY ISSUES

For many, anonymity is an essential precondition of gamete donation. For those who would seek to encourage (or at least not discourage) gamete donation, the logical approach would be to afford absolute confidentiality.[31] They would argue further that, apart from the practical benefits, affording donors confidentiality amounts to no more than an extension of a central principle of health care provision – that of medical confidentiality – to a new area of medical practice.[32] Compelling as these arguments may seem, reflection unearths two equally powerful countervailing considerations which suggest that the right to confidentiality should be, at most, a qualified right and perhaps even denied altogether. First, paternal anonymity has the potential to cause psychological harm to the child; especially where the child grows up fatherless.[33] So long as absolute confidentiality is afforded to non-social, genetic parents, the risk of psychological damage from growing up fatherless is thought by some to be unavoidable.[34] A precisely corresponding argument can be made with respect to ova donation in relation to motherhood. And yet, an insistence on the anonymity of gamete donors obviates the other obvious problem of the child growing up, seemingly, with two mothers or two fathers.[35] Nonetheless, it is submitted that it is more harmful to grow up with no mother or father than with two mothers or fathers. Certainly, step-children often happily adapt to more than two adults fulfilling the role of mother or father.

The second anti-confidentiality argument stems from Article 8 of the European Convention on Human Rights and Fundamental Freedoms, providing respect for citizens' private lives. In interpreting Article 8, the European Court of Human Rights has stated that it embraces the right that 'everyone should be able to establish details of their identity as individual human beings'.[36] This means there is a right on behalf of an assisted conception child to discover information relating to his or her genetic heritage. If this principle is not to be abrogated, it follows that a right for gamete donors to have nothing about themselves revealed to children who are genetically theirs must be denied.

In the light of the foregoing, it is not surprising to find that the Human Fertilisation and Embryology Act 1990 attempts to effect a half-way house solution to the confidentiality issue. It affords an 18 year old, born via assisted conception, the right to seek only a certain degree of information concerning his or her genetic parents from the Human Fertilisation and Embryology Authority.[37] The Act specifies that the Authority must furnish the applicant only with information that would disclose whether a person that the applicant proposes to marry would, but for the status provisions of the Act, be related to the applicant.[38] In addition, an applicant can obtain a limited amount of information about the gamete donor, as of right, under Article 8 of the Human Rights Convention.[39] Such information will not, however, for the reasons set out earlier, reveal the identity of the gamete donor.[40] Instead, it will be confined to details about the donor's educational background, skills, physical characteristics and health history. The value of such limited information to someone concerned about their very identity is highly questionable. Whether their curiosity will be assuaged by learning such trivial details must be doubted. At least, so much has been argued in relation to the corresponding case of an adopted child.[41]

In practical terms the English approach is probably as near to an acceptable, pragmatic compromise as one can come. It preserves anonymity in that the name and address of one's genetic parent cannot be obtained. It therefore fails to inhibit gamete donation. On the other hand, it does allow children of the new reproductive era to discover potentially vital information about their genetic parent's health history. A more open model of regulation, affording no anonymity, would have meant that a whole host of infertile and childless would-be parents would have been the ultimate losers. Equally, to have afforded absolute confidentiality would have involved running the risk, albeit small, of inter-breeding.

CIVIL LIABILITY FOR MISHAPS IN ASSISTED CONCEPTION

Because medically assisted conception is very much in its infancy, and because the techniques currently in use stand in need of considerable refinement, it is not uncommon for the children born to suffer from physical or mental abnormalities. Moreover, since the provision of IVF and other such treatments usually involve huge expense to the recipients, there is very real chance that they will seek to sue those whom they perceive to have been at fault for the disabilities that manifest in their children. Here, the problem for the law is whether to award (and if so, how to quantify) monetary damages in connection with those disabilities.[42] One problem is that to do so amounts, arguably, to the (wholly inappropriate) commodification of human deformities (see Duxbury, 1996, for an opposing view). An example explains the point. If we were to place a monetary value on suffering from Down's syndrome (i.e. the measure of damages), whilst we compensate the 'victim' we also, indirectly, place a derogatory label upon all other Down's syndrome sufferers.

Assuming the right to sue in respect of disabilities caused by the negligent provision of treatment services,[43] a second problem lies in identifying who possesses that right. Should it lie with the disappointed parents or with the child? In relation to this problem the English courts have distinguished two separate legal actions: 'wrongful birth' and 'wrongful life'. The former represents a lawsuit brought by the parents. The gist of the action is that the child is of a kind that was never planned or desired and that, further, it would never have been born but for the defendant's negligence. By contrast, an action for wrongful life is brought in the child's own name. The claim is that the child should never have been born with the disabilities from which she or he suffers and that it would have been better never to have been born at all. In effect, the child claims that because of the negligence of the defendant, the parents were never offered the opportunity to have the pregnancy terminated.

Of the two actions, the English courts appear prepared to tolerate wrongful birth claims but not those for wrongful life.[44] The rationale for this approach is captured succinctly in the judgment of Lord Justice

Stephenson in *McKay* v *Essex Area Health Authority*.[45] There, his Lordship opined that allowing an action for wrongful life:

> would mean regarding the life of a handicapped child as not only less valuable than the life of a normal child, but so much less valuable that it was not worth preserving ... that a child has a right to be born whole or not at all, to be born perfect or 'normal', whatever that may mean.[46]

Despite the reluctance at common law to tolerate an action for wrongful life, an IVF child's right to sue in respect of negligently inflicted disabilities has been created by the 1990 Act: it extended the remit of the Congenital Disabilities Act 1976[47] to cover children born by way of assisted conception. Section 1A of the 1976 Act now provides a right of action where it can be shown that the pregnancy was brought about by 'treatment services', that the disability resulted from a negligent act or omission in selecting, keeping or using the embryo or gametes, and that neither parent knew at the time of the provision of those services of the risk that the child could be born with a disability. Undoubtedly, it is difficult to reconcile section 1A with the common law rules since it represents an inroad into the principle that wrongful life actions will not be tolerated under English law. The Act, by sleight of the draftsman's pen, does this by making the action available in relation to the child's disabilities rather than by making it referable to the absence of a termination. But this is a distinction without substance. It is one that is more apparent than real for the very terms on which the action exists make it clear that it only lies because the person providing the treatment was at fault in using a defective embryo or gamete. As such, the claim is effectively that the child had 'a right to be born whole or not at all, to be born perfect or "normal"'; the very right that Lord Justice Stephenson was unprepared to countenance.

As regards wrongful birth, the question is how, and on what basis, such an action may be brought in the context of assisted conception. The answer derives both from rules of common law and statute. First, under the common law, an action may be brought (assuming that there has been payment for the treatment services) in accordance with principles of the law of contract: the birth of a handicapped child does not conform with what the patient expected and paid for. Thus, if she can show that those services were not supplied with reasonable care and skill, she may sue.[48] Second, for those who do not pay for treatment services, the general tort principles expounded in the *McKay* case can be applied. The action simply requires the plaintiff to establish that the child's disability or deformity was caused by the negligence of the person who provided the treatment services. This, however, sounds beguilingly simple. Establishing proof of what caused a disability is by no means easy, especially for persons without any medical expertise, since proving the pre-natal aetiology of human disability requires esoteric knowledge.[49]

Though some might find repugnant the notion that a civil law action for damages should be made available in relation to disabilities caused to children born by use of the new reproductive technologies, it is nonetheless the case that potential liability helps to ensure that those who provide such treatment do so with proper care. The possibility of being found liable in damages does much to guarantee professionalism in treatment provision. However, it is difficult to deny that allowing lawsuits for negligently performed assisted conception has the incidental effect of commodifying human reproduction. This result is difficult to square with the principle underpinning the Surrogacy Arrangements Act 1985, that commercialised reproduction is abhorrent and to be proscribed.[50] Furthermore, it is arguable that civil actions are an unnecessary means of ensuring professionalism since alternative methods are already built into the 1990 Act. For example, treatment services may only be provided lawfully by a person holding a licence granted by the Licence Committee of the Human Fertilisation and Embryology Authority.[51] Equally, in addition to requiring the award of a licence, the Act also requires the regular policing of licence holders: 'a Licence Committee shall arrange for any premises to which a licence relates to be inspected on its behalf once every calendar year, and for a report on the inspection to be made to it.'[52] Such periodic review, coupled to the power of the Licence Committee to revoke or suspend the licence, might be seen as sufficient guarantors of the professional and safe performance of treatment services. It is thus difficult to justify the creation of a statutory wrongful life action under section 1A of the Congenital Disabilities Act 1976.

EMBRYO RESEARCH

Often, people are able to answer affirmatively the broad question of whether the law should permit biotechnology to provide a partial solution to the problem of infertility.[53] What they frequently find more difficult, however, is the question of what ought to be done with any surplus embryos, after a successful cycle of IVF has achieved a pregnancy. Should such embryos be allowed to be used for the purposes of medical research (and even then, within what constraints)? Or should we simply flush them away (or is even this morally and legally objectionable)? The answer to these questions turns, for many, on the moral status that we attach to the human embryo.[54] If it is as much a human being as a fully grown adult, then the idea that it should be subjected to destructive scientific research is unthinkable. Others, however, would argue that the absence of pain (regardless of the embryo's moral status) is the key to whether research should be permissible. Indeed, for them, the absence of pain makes it more acceptable to conduct research on embryos than sentient animals, such as rats. Still others would argue that the alternative – flushing the spare embryos away – produces exactly the same destructive result as performing research thus justifying research on basic utilitarian principles.

In truth, the debate about the propriety (or otherwise) of embryo research is all but intractable and there is insufficient space here to rehearse in depth the various antithetical arguments. Suffice to say, if there is a universally acceptable answer to this question, it is buried deep in notions of personhood, ethics, morality, religion and pragmatism.[55] Nonetheless, the intractability of the problem provides no reason why the law should ignore the acute question of what can and should be done with spare embryos. English law, arguably undecidedly, has endeavoured to steer a pragmatic, middle course.[56] First, it prohibits any research beyond the point at which the 'primitive streak' appears in the embryo. This occurs at about 14 days and marks the point at which embryonic matter becomes potential human life. Secondly, the 1990 Act also prohibits the use of human gametes and embryos in any attempt to breed across two species.[57] Third, subject to these prohibitions, the Act does permit embryo research for a specified, limited series of purposes. These include promoting advances in infertility treatment, increasing knowledge about the causes of congenital disease, developing more effective methods of contraception and discovering methods of identifying gene or chromosome abnormalities.[58] In this way, English law seeks to recognise the 'wasted potential' argument: that simply flushing spare embryos away is no more morally repugnant than using those embryos for the common good of humanity and future generations. On the other hand, the legislation avoids the worst-case scenario of research for research's sake. In utilitarian terms, the Act probably fails to maximise satisfaction, but it probably also succeeds in minimising the dissatisfaction that is inevitable whenever spare embryos are produced. In a context in which legislation is indubitably required, and in which it is impossible to please all-comers, the Human Fertilisation and Embryology Act probably strikes as acceptable a balance as is possible.

THE REGULATORY FRAMEWORK

So far, I have attempted to demonstrate that regardless of whether one approves of assisted conception by recourse to reproductive technologies, the law is forced to supply some form of regulation. The very fact that it is possible to achieve a 'test-tube' birth demands at least some legal response (if only to resolve the status issues). Not everything the scientists achieve will, or should, be legally permissible. And if the law fails to state that procedure X is only permissible in certain circumstances, or that it is never permissible, it impliedly states that procedure X is *always* permissible. Given this necessity for a legal response, I have attempted, by sketching the *substance* of the English legal position, to illustrate the various arguments both for and against the kinds of rules that might be settled upon. What I have yet to do, however, is to say something about the *form* of regulation adopted. I have alluded to the existence of a licensing system but said very little about the way in which that system is administered. It is to this final issue – the regulatory framework – that I now turn.

The 1990 Act made provision for the creation of the Human Fertilisation and Embryology Authority. This authority is responsible for issuing licences to provide assisted conception and to conduct those forms of embryo research permitted under the Act. The Authority, in carrying out these functions, is not required to administer a strictly defined code of rules but is instead invested with a very broad discretion as to how they are performed. By this means Parliament has, to an extent, 'passed the buck' as regards defining who may and may not provide assisted conception services. Likewise, the fact that the Authority is ultimately responsible for issuing research licences means that the UK Parliament is not directly responsible for authorising projects or experiments that go right to the very frontiers of ethical acceptability. The only check placed upon the Authority's conduct of its prescribed functions is that it must file annual reports with the Secretary of State detailing its activities over the last 12 months and those planned for the forthcoming 12 months.[59] Outside of this minimal level of accountability the Authority has more or less complete control over health practitioners and research scientists.[60]

Since it was the clear intention of Parliament that regulation of assisted conception and embryo research should be regulated by the Authority, it is of obvious importance that its membership should comprise respectable, knowledgeable and responsible members of society. Schedule 1 of the 1990 Act governs the issue of membership. It specifies that at least one-third, but fewer than half, of the members should be registered medical practitioners or embryologists.[61] Outside this 'power bloc', however, there is no restriction on who may be a member other than that they must secure their appointment from the Secretary of State.[62] In this way, the Authority's membership assuredly extends to leading public figures, drawn from the ranks of 'The Great and the Good', and not merely medics. It embraces eminent lawyers, sociologists and high-profile religious personalities.

Whether the composition of the Authority with its limited amount of 'public representation' ensures that treatment and research licences are issued according to moral or ethical principles is doubtful. The debate that takes place prior to the award of a licence will inevitably be informed on a scientific level given the presence of the medics. However, there is a danger that their scientific expertise will dominate to the extent that the scientific merit of a proposal, backed *en masse*, will drown out any arguments advanced by a solitary theologian or ethicist.[63] Equally, since the non-medics and non-scientists have other commitments beyond their membership of the Authority – it not being a full-time appointment – it is to be doubted whether, by the time their membership expires, they will ever have gained enough esoteric knowledge to discuss such matters as confidently and convincingly as the scientists and doctors.

A further limit on the effectiveness of the 'public' representation in the Human Fertilisation and Embryology Authority is the fact that its members are to be selected and appointed by the Secretary of State.[64] To this extent, rather than being a truly independent and apolitical body – akin to a research ethics committee, for example – the Authority is instead necessarily a political assembly.[65] A still further constraint on the effectiveness of the Authority as a mean of regulation is the fact that its decisions about who should and who should not be granted licences are not something often found in the public domain. Consequently, the decisions of the Authority cannot be policed in the usual way. How do the uninitiated police force know whether X had a licence or not and whether X was breaking the law in conducting his research project? What is perhaps needed is a Human Fertilisation and Embryology Executive.[66]

CONCLUSION

What we have seen is that the law has had to respond to a number of challenges spawned by the reproductive revolution which are both unfamiliar in kind and especially difficult to legislate upon. It would be surprising indeed if a simple statutory code could provide a universally acceptable set of solutions to each of these problems at a single stroke. Giving the

Authority the power to create and amend its own Code of Practice recognises the practical need for flexibility both in the rules themselves and in the regulatory framework. The model of regulation that was adopted in Britain also has the advantage of taking a number of intractable medico-moral issues out of the hands of Parliamentarians (and hence out of the realm of politics). Instead, and much more appropriately, it entrusts the relevant rule-making to those professionally concerned with assisted conception and embryo research. Creating legislation that is a panacea is an impossible dream. The Human Fertilisation and Embryology Act recognises that impossibility and attempts to achieve of broadly acceptable, pragmatic solution to a host of late twentieth-century problems.

8 Discourse Formations and Constellations of Conflict: Problems of Public Participation in the German Debate on Genetically Altered Plants

Alfons Bora

THE LIMITS OF PARTICIPATION

One way for society to respond to innovations in biotechnology is to involve the public as broadly as possible in relevant decisions and processes. In a certain way this approach has been the traditional answer that modern democracies give to fundamental conflicts, for public participation is an aspect of democratic decision-making processes. It is called for and practised especially when alternatives for exerting control and influence appear to fall short. That is often the case when particularly drastic or far-reaching consequences are imputed to decisions, such as those on introducing new technologies whose social and environmental impacts are still difficult to assess. Such circumstances in particular also spawn the distrust of experts. And expertocracy can be opposed only by the active involvement of laypersons – popularly known as 'those concerned' – in socially relevant processes of evaluation and decision-making. At least that is the call in many quarters. For this reason, grass-roots participation and round-tables are widely considered a means for successfully managing conflicts and problems and, hence, as promising instruments of social integration.

The range of functions attributed to public participation varies. The optimistic standpoint that public communication is a form of collective problem-solving was standard in pragmatism (Dewey, 1927) and the Chicago School (Blumer, 1966). Today, however, it lingers, at most, in the political philosophy of communitarianism. The assessment is much more reserved in Critical Theory (Habermas, 1996), in which the autonomous public is ascribed a rather indirect influence on formal decision-making processes and posits that the public's main function is to legitimate deliberative policy.

Within the legal system, forms of participation are primarily intended to optimise the exercise of individual rights. The rights to be heard and to participate ensure legal protection early on and are intended to help control administrative action. In this sphere participation amounts primarily to control over procedures, whereas control over results remains with the legally institutionalised organs (courts and public administrations).

Grass-roots participation has become increasingly controversial, however. It is not only critical thinkers within the academic community who are diagnosing that participatory democracy has broken down. Indeed, it is more striking that a great deal of empirical evidence in support of such assertions seems to be coming from environmental associations. Walkouts from 'round-tables', technical impact assessments and other kinds of grass-roots participation are unmistakably mounting.[1]

Where does the problem with public participation lie? One source is bound to be the peculiarities of Germany's political culture, which is especially polarised over the issue of genetic engineering. These rifts do not exactly improve conditions for processes that could forge public consensus.[2] Of course, that explanation does not go far enough given the fact that most non-governmental organisations (NGOs) continue to urge participation. A more likely reason for the trouble thus seems to be the forms and consequences of the participation processes themselves.

In the following pages, I will therefore study the question of which conflicts and fault lines originate in the clash between the different 'language-games', the discursive formations, of the communication process. Space restricts me to an examination of public participation in legally mandated decision-making processes.[3] The hypothesis I defend is that public participation in legal decisions can be justified as compensation for the absence of 'externalisation' of responsibility for dealing with risk from the legal system to the decision-making community, science and business. At the same time, however, this participation tends to blur or eliminate social differentiation, an effect resisted by the legal system. This argument is developed in three steps. I first describe how risky decisions are passed back and forth between the policy-making, scientific and legal communities, and which mechanisms the legal system develops to deal with risks. I then discuss the conflicts that have arisen within the legal system over genetic engineering in agriculture. I conclude by developing a heuristic tool based on conflict theory, a tool that allows the dispute over biotechnology to be decoded as the tension between diverging formations of discourse.

RISK: PASSING THE BUCK IN THE DECISION-MAKING, SCIENTIFIC AND LEGAL COMMUNITIES

The modern legal system is full of risks. The number of hazardous decisions it is called upon to make are multiplying because it has partly departed from its normal if ... then programme and thus no longer ensures that responsibility for risk is shifted to other levels, such as the decision-making community, science or business. As legal codes become increasingly responsive to ecological concerns, public administration acquires more and more planning and designing authority that expand the execution of legal programmes into a responsibility encompassing the aspect of standard-setting as well. From a sociological perspective, it is now possible to see that this growth in authority and responsibility is involving the legal system in a growing number of risky decisions. This development is illustrated and its impacts examined in the following pages. A brief discussion of sociological aspects of the risk concept prepares the ground for exploring the view that shifting risky, technology-related decisions from the policy-making and scientific communities to the legal system typically leads to demands for public participation.

Risk and Decision-making

The concept of risk depends on one's theoretical assumptions. The behavioural sciences have long pointed out that the 'formal, normative' concept of risk as developed in shipping and the insurance industry, involves certain problems. Expected loss per unit of time (amount of damage multiplied by the probability of its occurrence) cannot be established as a uniform measure of risk. For one thing, it is usually difficult to quantify benefits and damages on one scale in order to compare them. For another, there is often no criterion by which to determine the extent of damage. The cases that are socially the most contentious are precisely the ones in which no empirical basis exists for calculating probabilities. Lastly, many studies show that the commercial calculation of risk typically diverges from the risk perceptions of everyday reality and the body politic. That is why cognitive approaches to risk research emphasise the individual and collective perception and evaluation of risks. Researchers taking cognitive approaches study such questions as what the willingness to engage in hazardous behaviour is a function of and seek the categories according to which responsibility is attributed.

Sociological risk research focuses on social processes of communication from which such semantics of risk arise. Trying to illuminate the social conditions governing opportunities for discussion, scientists in this area of endeavour show which concepts of risk are particularly popular on which occasions and what the influences are that promote this cycle. Douglas and Wildavsky pointed to the social construction of risks. In coining the term 'risk society', Beck (1992) made a major contribution to today's discussion of 'environmental problems' as society's problems.

Modern systems theory directs attention to the relation between risk and decision-making. Risky, or hazardous, decisions are described as the choice between alternatives that may have adverse consequences for third parties and that then must be answered for to them (Luhmann, 1993b). Conceptually, the risk of making a decision is distinguished from the danger of damage. Whether something is thought of as a risk or as a hazard depends, first, on whether it is attributable to someone (see also Douglas, 1990). Unlike natural disasters, for instance, risk is thus a socially self-created opportunity to inflict damage. This differentiation not only calls attention to a trend in the historical distribution of hazards and risks – the fact that more and more situations are being transformed from a hazard into a risk, that the number of society's decisions, and thus the degree of contingency, is growing. It also reflects the fact that the relation to the future has changed in modern societies. Impacts lying in the increasingly remote future must be considered today. The consequences of taking action or of doing nothing are often equally complex and unpredictable. Ultimately, this circumstance ushers in a distinction characteristic for due process: that between the decision-makers and those who may bear the brunt of the hazard. The differentiation between the risk entailed by society's decisions and a possible danger coincides with new differentiations within society itself.

What complications arise in principle when hazardous decisions are to be made? Very generally, decision-making entails a variety of typical dilemmas and paradoxes. One of them is the control dilemma, to which Collingridge (1980) pointed: at the point when it is still possible to control the inception of a new technology, human technical expertise far outstrips knowledge about that technology's social impacts. The social impacts cannot be forecast at that point. By the time they have been discovered, the technology is so firmly entrenched in society that controlling it is practically no longer possible, or only very slowly or at great cost. This dilemma brings up the matter of the time structure involved in hazardous decisions. Although the time structure itself harbours no paradox, it suggests a crucial aspect called risk paradox – the constitutive connection between the pressure to make a decision and the impossibility of knowing the ramifications of that decision at the time it must be made (Luhmann, 1993b). To this aspect of uncertainty is added what Clausen and Dombrowsky (1984) characterised as a warning paradox. It reminds one that warnings of danger cannot in principle make hazardous decisions easier. One cannot tell if a warning was justified unless one *ignores* it. By heeding the warning, one will never know if it was warranted.

ASSUMING RISK AND SHIFTING RESPONSIBILITY

Against this background, the question arises as to whether legal pro- grammes exist for hazardous decision-making. They do for classical if … then programming. The legal system can deal with such risk decisions. It normally has various types of programme at hand. It may proceed as it would when judging risk decisions made by third parties, in which case it acquires from other areas of society (science or business) certain criteria stipulating what behavior is to be coded as permissible or impermissible. Examples of this programme are minimum and maximum limits and the practice of making a legal decision contingent on the ever-changing scientific state of the art. Or the legal system takes the vicissitudes of deci- sion-making into account by providing a degree of latitude for choosing the legal course of action to be pursued after the operative facts of the case have been ascertained or by building in time-frames within which those facts are considered relevant.

A characteristic shared by the 'classical' programmes is their tendency to shift responsibility for hazardous decisions from the legal system to other parts of the social system. Maximum and minimum limits and the choice to make legal decisions depend on the constantly changing scientific state of the art are mechanisms to ensure that the scientific community ultimately bears the risk of being made responsible for errors and faulty forecasts. The legal system merely executes the scientific com- munity's decisions. Whatever lies below the scientifically or politically established limit is considered legally permissible.

In the case of new technologies, however, this mechanism is interrupted when the legal system is abandoned by the policy-making, scientific and business communities, when it is left to bear decision-making risk alone. It is true today particularly when extra-legal processes known as standard-setting fail, or fail to be accepted. Many instances of such failure have been documented. The policy-making community is often no longer in a position to decide on the acceptability of technology and put it into legal form. The scientific community is being made aware by internal criticism that its causal models have been too simple, that it has been dealing with idiosyncratic processes where technological developments are concerned. Minimal and maximal limits are understood as 'political' limits. Protest movements are forming whose 'semantic politics' (van den Daele, 1990, pp. 20–1) ensure that the subject of 'technology's risks' does not disappear from the public agenda. The business community refuses to accept sole liability for what may be unlimited damage if technology as a whole is considered by society to be worth promoting.

One should therefore speak of having the legal system make risk-related decisions in those cases where extra-legal criteria do not provide a complete, or any, foundation for them. In some cases of dealing with new technologies, I presume that the legal system will be compelled to assume precisely this responsibility. It loses the innocence of the 'neutral' by-stander. The risk upon which the legal system enters in this regard is that it may help cause damage because of the self-programming (standard-setting) that precedes the legal decision and that it may have to justify to the victim(s) the choice it made between discretionary options. Normally, the if ... then programme of law provides an adequate basis for justifying that choice. But in some cases where modern high technologies are involved, these mechanisms for shifting responsibility break down for various reasons.

From the sociological perspective, the developmental trends of law governing environmental administration and technology can thus be interpreted as legal consequences of the development toward the risk society in a very specific sense – as a response to the absence of inputs from external programming. In this situation, the obvious thing is to at least gain the consent of the parties that could be affected. For new technologies whose impacts are difficult to contain over time and space, that approach would mean allowing *every* person to participate in the actual process of setting standards and making decisions in order to identify possible dangers, control procedures and clarify the licensing issues. This public participation does not change the uncertainty of the decision, but it does compensate for the failure of the policy-making community, which has failed to forge social consensus on the issues involved with new technology and then left the legal system standing alone with decisions on whether to permit it. Public participation, at least in theory, broadly distributes causal attribution and responsibility for adverse impacts. It takes pressure off decision-making organs while promoting the willingness of the participants to accept it by virtue of their involvement.[4]

Therefore, if the legal system inherits risk-related decision-making responsibility that overburdens its customary programmes because that responsibility cannot be passed on to other parts of the social system, if, in other words, a need for programming (i.e. standard-setting) arises, then public participation in the form of everyone's involvement shall compensate for this shortcoming. The postulate of public participation can therefore be reconstructed as an attempt to legitimate risk-related legal decisions in the absence of the inputs needed for making those decisions in a functionally differentiated society.

THE DILEMMA OF PARTICIPATION: AN EXAMPLE

Thus far I have illuminated the legal and sociological background of public participation in environmentally relevant licensing procedures. In the following steps of argumentation, the problems arising from such forms of participation are considered. They are manifested primarily at the level of communication in public hearings. However, they can be explained only with an eye to the features characterising system-specific codes that collide in these communications.

Functional differentiation means, among other things, that the different parts of the social system use their own codes. In other words, they operate on the basis of their own differentiations, and only them. In the legal system, this circumstance is seen, for example, in the fact that the linking of decisions to the code of right and wrong essentially bans all other options (e.g. wealth, power or love). This stipulation of a binary code serves as the anchor for, among other things, the social functions of law. To perpetuate the expectation of a given behaviour, it is necessary to ensure that the consequences of a legal decision will be based on the differentiation between right and wrong. Analogously, the legitimation of the decision depends on the possibility of checking its correctness, and so forth. However, binary coding is not linked to a codification. Theory of the sources of law also allows for tacit legal custom, suprapositive principles or judicial self-restraint. The important thing is to protect the legal system's identity of the system–internal stability, which is understood to mean consistency (uniformity of programming and application of the law) and continuity (predictability of decisions, law and order). Only through such structural stability is it possible to distinguish between other systems and their codes. That is, only then can there be social differentiation.

The purposes of the observations discussed in the following pages is to support the assumption that public participation tends to bring about disruptive differentiation effects within the legal system at the level of communication in public hearings. For in those kinds of communication, especially in the cases cited here, the law is confronted by certain expectations and claims of a non-legal nature that have a direct bearing on decision-making. It is presumed that the system preserves its autonomy by attempting to fend off such menacing interventions. But such response reduces the willingness to assume responsibility for legal decisions.

Until 31 December 1993, section 18 of the Genetic Engineering Act 1990 mandated an oral public hearing in particular circumstances, especially when petitions for the release of a genetically modified organism were involved. In such a case, every person – hence, every person in the entire country – was permitted to raise written objections. As the licensing authority, the German Federal Board of Health invited all objectors to an oral discussion, at which all the arguments that had been raised had to be considered. Although persons other than the objectors were not allowed to participate in principle, anyone wishing to be present was admitted. In these kinds of proceedings, objections must always relate to the risks and hazards of the genetically modified plants. There is no control over the decision by vote on acceptance or rejection, but the right to raise factual rebuttals provides procedural control basically enabling opponents to exert real influence on the result of the proceedings (Bora, 1994).

In principle, one can imagine at least five kinds of arguments that could confront decision-making bodies in such situations: (a) scientific, (b) legal (in the sense of substantive law), (c) procedural, (d) substitutive, and (e) methodological-evaluative. Scientific, substantial and procedural aspects pose no obstacles to the legal system and are routinely handled through standard rules of application. For the enforcement and interpretation of law, cognitive issues play a key role when it comes to establishing the operative facts of the case. At that point experts disagree on what is to be considered a 'case'. This disagreement is the point of departure for the legal evaluation, which also deals with all conceivable material and procedural issues in the interpretation of existing law. Only the substitutive and methodological-evaluative kinds of argument are to be considered problematic. They are examined below, each in turn.

One point must be made first, however. When 'problem zones' of public hearings are treated in the following passages, it needs to be clear that many of the debates at these events about genetic engineering and its possible dangers have been completely unproblematic. They have mainly involved disputes about the first three types of argument cited above. The fact that only the negative sides come out in this article is due to the topic of this article: the issue that the adverse effects have ultimately contributed to the failure of public participation and its general deletion from the Genetic Engineering Act 1990. The causes of this failure should therefore be of interest.

Substitution of the Legal Code

Experience has shown that moral valuations, political preferences and numerous other viewpoints are discussed at the public hearing. All these positions claim to provide the legal system with relevant criteria on which it must directly base its decisions. The substitutive type of argument as a special case can be recognised by its attempt to replace the legal code of 'right and wrong' with other criteria, such as morals or politics. If this intervention were to succeed, it would trigger a trend towards blurring or eliminating social differentiation, an eventuality to which the system regularly responds by closing itself to all such endeavours, a result perceived as delegitimating in the legal system's context. As will be seen, one of the standard reasons for closure is the fact that bringing non-legal arguments into the legal discourse reveals its basic paradox. Closure, then, serves the purpose of mitigating, eliminating or obscuring the paradox.

Such attacks assume different guises. In public hearings called in accordance with the Genetic Engineering Act 1990 one finds a positive law critique with a rationale approaching that of natural law and with a direct appeal to the discretionary bodies not to adhere to the Genetic Engineering Act. It is said, for instance, that there is a 'natural law' that exists in harmony with the 'real laws of human rights' and that it takes 'common sense' and 'ethics' to obey this natural law.[5] Whoever recognises this viewpoint immediately sees that the Genetic Engineering Act 1990 is simply 'wrong'. It is pointed out that 'the greatest crimes against law and order' are being committed. For that reason, the appeal is aimed at the 'conscience' of those people representing the authorities, to their 'responsibility as human beings'. Under these conditions, the Genetic Engineering Act is 'not right and we mustn't adhere to it'.

It ought to be clear that these arguments are not directly relevant to the application of law – for very much the same reason that, for example, one cannot in principle justify a decision by saying that a majority of the objectors who were present wanted it that way. One can cite two reasons that such types of rationale are generally rejected on the grounds of existing law.

1. Such interventions within the framework of public participation are a threat to the structural identity of the legal system. To the extent that they appeal directly to the decision-making bodies to exceed the legal code, they differ from the above-described mechanisms for adopting outside codes (e.g. those of the scientific, policy-making or business communities) in the legal system, for the intention is to replace the legal code's rules with others. It is no longer the law that determines which external tenets become relevant under which conditions. Instead, external codes (particularly morality and politics) tend to stipulate the programming at this point.

From the perspective of the legal system, however, this change is allowed only in legally sanctioned procedures (legislative ones, for example). For such structural interventions compromise the ability of the system to function in a complex environment. For every system, the relation between system and environment expresses a complexity differential (Luhmann, 1995, pp. 45–51). The system must have the requisite variety in its relation to the environment (Ashby, 1956, pp. 206–8) in order to respond to the environment's complexity and generate in other systems a willingness to accept the legal system's outputs. If the structures of the legal system are adversely affected by non-legal codes, the system's complexity threatens to fall so low that the difference between it and that of the environment becomes great enough to threaten the existence of the system.

2. The kind of interventions in the structure of law that have been outlined here are especially precarious. They clearly show the paradox that underlies the distinction between right and wrong and every other differentiation, the paradox that the legal system keeps invisible in its routine operations (Luhmann, 1993a). This paradox is that no distinction can be applied to itself. It is impossible to tell, for instance, whether it is right to distinguish between right and wrong. The 'origin' of the distinction always remains the odd man out. In the case of the legal system, this third party is the system's power-related origin.[6] And the validity of the right/wrong binary code is directly challenged by public participation. In the call to ignore the law, the non-legal origin of law comes out. Because a supra-positive right to resist cannot be inferred from law itself, such interventions are ultimately doomed in legal discourse. They spark no response and are regularly dismissed in the hearing on the grounds of 'law and order' or 'existing law'.

There are thus two motives militating against the direct validity of outside codes in legal proceedings: (a) preservation of identity and the development of complexity, and (b) concealment of the legal system's basic paradox. For that reason, the emergence of regulations that represent an attempt to thwart such external interventions in the autonomy of the legal system has been utterly functional in terms of perpetuating the legal system's structure. The primacy of the law, the existence of procedural rules and the delineation of authority ensure that the communication of the legal system cannot be randomly replaced by some other and that reference to extra-legal operative facts of a given case is regulated according to the legal system's own criteria. It is obvious to presume that this stance, in turn, causes unease in political and moral discourses.

Conflicts of Application: Criteria and Methods of Evaluation

The second type of argument was introduced above under the name 'methodological-evaluative'. In public hearings there are a great many controversies surrounding the general principles on which an evaluative or subsumption procedure should be based. Ethical principles such as the protection of life and health, nature and the environment, and future generations are not at issue. As abstractions, these points of view enjoy broad consensus. The concern is rather with questions having to do with applying such principles in a specific case. Consider the following four topics, for example:

1. *Hypothetical (Speculative) Risk*
In the dispute over new technology, specifically genetic engineering, a special type of cognitive uncertainty plays a prominent role. Whereas the classic legal distinction between hazard, risk and residual or marginal risk still begins with the assumption that causalities (possible paths of causation) are recognisable and that at least the magnitudes of probability that they will occur are ascertainable, these essentials do not apply to speculative risk. Speculative risk differs qualitatively from the other forms of risk in that one does not know which dangerous event could take place at all, much less how probable it is or what the possible damage could consist of. Usually, the argument is that the 'novelty' of genetic engineering eliminates the possibility of ruling out some kind of catastrophic damage.

'Hypothetical (speculative) risk', then, does not mean all those cases in which the presumptions of safety that are based on hitherto uneventful operation are shaken by reference to data and publications. Precisely that kind of approach would be the standard, legally acceptable sort of reasoning. The logic of the argument in the present instance is different, however. In the first step, indeterminism – the observation of nature as an idiosyncratic system, the experience with past accidents – is used to underscore the general unpredictability of a technical innovation's possible consequences. In the second step, the conclusion is drawn that one can in principle never preclude the possibility that something catastrophic will happen when an innovation is involved. From this statement, which is indisputably true, it is then inferred that a license for genetic engineering is to be refused in each specific case.

However, this argument triggers a defensive response in the legal system, for it is tantamount to reversing the burden of proof. For all its openness, the law governing technological safety continues to require at least the plausible presumption of risks. If uncertainty cannot be demonstrated at least 'in principle', safety is presumed. And this programming cannot be eliminated from the licensing procedure. Legal changes are possible only through legislation or, at very most, through judicial discretion. It is scarcely surprising, therefore, when 'hypothetical risk' is rejected on the grounds of existing rules governing the burden of proof. The fact that this response appears to make a mockery of the anxieties felt by potential victims is one of the frequently lamented results of such disputes.

2. *Criticism of Science*

A great deal of criticism erupts when it comes to the vague legal concept of the 'state of the art'. There is no legal problem with all those cases in which it is said that the data in the application for a license fail to meet to the state of the art in science and that this aspect must be improved.

It is quite a different matter to judge arguments opposing application of the legal entity itself. Such cases include positions in which 'self-restraint' in the licensing procedure is sought – a stance somewhat similar to that expressed in the hypothetical-risk viewpoint. The argument is that the state of the art in science is characterised by unpredictability and generally negative past experience, so no release of genetically modified material may take place or be allowed. A much more fundamental critique is heard as well, however. For example, it is said that referencing the legal system with the system of science is to be rejected in general because science does not treat the subject of its study 'as a whole', because science is not oriented to truth but rather to the interests of those 'whose hands it eats from'. Conversely, it is also asserted that modern science is in principle 'unreasonable' because it is interested only in truth, devoting itself to a disastrous 'separation between objectivity and ethical involvement', to a reification of the natural, and because it is living in a sinister coalition with a 'bureaucracy indurated by paragraphs'.

Such arguments go largely unheard in public hearings, however. In any case, they do not move the decision-makers to refuse the licence, for they ignore the legal distinction between giving and applying law.

3. *Assessment of Benefit*

Existing law says nothing about an *a priori* test of need for new technology. But precisely such a test is one of the things demanded in public hearings. It is argued that licensing of a technology for which there is no social need should not be permitted, quite apart from whatever risks there might be. According to this reasoning, it is necessary to check for benefit in any case because the risks are not known. The programme of law governing technological safety currently requires at least a suspicion of danger before public administrations can initiate a precautionary weighing of risks and benefits.

4. *Ethics of Assessment Criteria*

An example in this sphere is the call to make ethical considerations a direct decision-making criterion by interpreting them as part of the legal facts of the case. According to this viewpoint, 'socio-ethical' and 'economic and ethical' perspectives are to be included in the reasons stated for a decision. Furthermore, account must be taken of tendencies towards monopolisation and of the 'opportunities for rural agriculture'. This tack is usually met with silence in the public hearing or countered on the grounds of 'existing rules of interpretation'.

Using different examples, I have tried to illustrate how certain claims are defeated in the legal system. It is true that these claims do not end up replacing the code of right and wrong with other differentiations, but they could introduce inconsistencies and increased complexity. One might suppose that eventuality is precisely why the reactions are systematically defensive. Through extensive interpretation of the law, arguments of this sort are being used in an attempt to broaden the existing latitude for consideration of different standpoints. The endeavour is legitimate but is finding little favour among members of the legal profession. Some of the views raised could indeed be discussed and are perhaps politically insightful. Nevertheless, they have been rejected or, more often, met with silence presumably because the legal system, like all other systems, can afford only a limited degree of flexibility. Granted, existing legal solutions are not carved in granite, but they cannot be revised at will either. Reasons of internal consistency, namely, justice, compel this behaviour.

For law as a social system, there is in these respects always the danger of over-responding to environmental demands. Doing so would create problems of injustice within the system. The reason is that complexity, as far as legal issues are concerned, is not only an external issue of adapting to the environment but also an internal issue of systemic development. In the legal system the concept of justice has the function of signalling 'adequate complexity', meaning the maximal degree of complexity that can be tolerated without losing the internal consistency of the system (Luhmann, 1981, p. 390). If complexity exceeds that point, it is experienced as inconsistency and/or discontinuity, that is, as a violation of the principle of justice. In that case, the only external viewpoints that can be readily accommodated in the process of risk-related legal decision-making are obviously those that can be integrated into the legal system in a manner consistent with its structures. Claims and demands that go beyond that point are likely to be rejected on the grounds that they are incompatible with the system.

CLASHING DISCURSIVE FORMATIONS – A HEURISTIC TOOL BASED ON CONFLICT THEORY

At the socio-theoretical level, this reconstruction is still a rather sketchy description of the relations between the legal system and its environment when there are risky decisions to make. A more detailed study of the matter would have to be able to illustrate such blurring effects at the theoretical level of interaction as well. The empirical analyses would require conceptual tools with which to show how the theoretically presumed functional shortcomings of public participation are actually manifested in the tension between different manners of perception and types of communication at actual public hearings.

For this reason, a few concluding comments about a theoretical concept of discourse are in order, beginning with the concept of 'discourse formation'. Discourse formation is understood in this context as an ensemble of implicit rules underlying a discourse. Discourse is understood to be a coherent complex of communications, but not necessarily one that forms a system yet (in the sense found in autopoietic systems theory, for instance).

Not every context of elements constitutes a system as well. That is why I speak of an 'ensemble' of rules. The term likewise indicates that a systemic connection need not be involved. Accordingly, the term 'rule' is modelled on the concept of structure. A rule, be it semantic or pragmatic, represents a constraint on the linkage between possible communications. If these rules are said to be 'implicit', it means that such discursive formations lie beyond the subjective consciousness of actors and are to be understood as objective entities.

The similarity between the term discourse formation and Bourdieu's concept of habitus is intentional but does not suffice for a theoretical definition. In his early definition oriented to Chomsky, Bourdieu understood habitus as a system of internalised patterns that allow the generation of all typical thoughts, perceptions and actions of a culture – and only that one (Bourdieu, 1974, p. 143). Just as habitus formations relate to social fields, the discourse formations in the present theoretical framework are embedded in contexts of social systems, without being identical to them. As with the concept of habitus, a trans-situational and trans-individual aspect is characteristic of discourse formation, lending it a certain life of its own (the 'hysteresis' effect).

However, reservations about Bourdieu's concept of class and about the body-oriented focus of habitus weigh heavily enough[7] to give priority to a conceptualisation centred on functional differentiation (instead of class antagonisms) and communications theory (instead of 'incorporation' of the habitus). These considerations are intended to come out in the term 'discourse'.

In the present context the concept of discourse is not used as it is in consensus theory (e.g. as represented by Habermas and Apel). It is more closely allied with semiological, structuralist and psychoanalytical traditions, which keep the concept of discourse normatively indeterminate. In this regard, it is closer to Michel Foucault's theory of difference and constructivist approaches. Foucault attempted to describe discourse as socially produced and controlled series of communicational events whose diversity represents the polycontextuality of different 'discourse communities' within a society. In 'The Order of Discourse', Foucault (1977) outlined this point, not only sketching out different mechanisms that preclude, discipline and otherwise constrain social discourse, but also illustrating how numerous discourses overlap within 'one' society.

My hypothesis is that this poly-contextuality is what shapes the dynamics of communication with regard to public participation in the legal system. As interfaces for different social subsystems and their competing rationalities (Bora and Döbert, 1993), public hearings would thereby be determined largely by the rivalry between discourses formed around the respective prevailing situational definition or type of rationality, that is, by a *conflict over the monopoly on interpretation.*

It is presumed that the event could be characterised by an intractable dispute about which rules are to govern communication, by a 'différend' (Lyotard, 1988). It can be presumed further that the unresolvability of this 'différend' is rooted in the fact that the conflict, on both sides, touches on the underlying paradox of the particular discourse, triggering allergic reactions as it were.

Representatives of public authorities insist on remaining within the legal bounds of decision-making. The rules of the legal discourse therefore determine the rights and responsibilities of the participants. Citizens appear, whether as members of social movements and political organisations or as members of the interested public. In either case, the logic of political discourse will compel them to transcend what from their perspective are the much too narrow constraints of the legal system and to confront it with the power of the law before the law (Derrida, 1994). Citizens will use arguments that challenge the validity of the legal system itself, will call for the burden of proof to be reversed and ethical considerations to be weighed, and may demand that the proceedings be terminated. In the course of such an encounter, complementary incompatibilities arise. One could say that the legal system remains blind to the fact that it is also power. Power remains blind to the fact that it communicates injustice under these circumstances. The blind spots lock the two discourses in a tug-of-war.

Theoretically, one may thus diagnose either that the structure of the legal system is threatening to slip below the lower threshold of requisite variety as a result of non-legal interventions – the danger of 'implosion' – or that it is being called upon to adapt its complexity to external demands and thereby raise the upper threshold of adequate complexity -– the danger of 'explosion'. Both effects can be avoided only if the legal system succeeds in balancing its structures stably between the two extremes. To do so, it will close its internal processes as much as possible to possibilities for intervention and tend to reject increases in complexity. That is why there is a trend towards pushing such changes onto legislative bodies, or in any case not establishing them in legal practice unless it has to, and even then only as inconspicuously as possible.

The consequences are predictable. Whoever hopes to expand the basis for legitimating risk-related legal decisions through direct and broad public participation is going to be disappointed, at least theoretically. In all likelihood, public participation in legal processes themselves will eventually undermine that legitimacy because within such proceedings all communication oriented to a code of law will tend to filter out other kinds of communication and exclude them from the decision-making process. If this theoretical characterisation should prove apt, it will be able to offer a sociological explanation of why public participation fails to relieve the legal system when risk-related decisions are involved.

9 Law and the Cultural Construction of Nature: Biotechnology and the European Legal Framework

Christian Byk

Everyone knows what a road is. It follows that everyone knows what law is: the line that divides the road down the middle creating lanes for circulation. Of course, everyone knows many types of road: most are two-way but some are one-way and you can have several lanes for each direction way. The line that separates the lane can be either a solid line or a dashed one. But in all cases, everyone knows that he/she has to drive to one side of the line – on the left or on the right according to the driving code of the country – but everyone also knows that if he/she is driving right on the line or zig-zagging across it, he/she would be breaking a legal rule. In this example, nature – by which I mean the landscape and the road – is not obviously the result of legal rules dealing with the best way of circulating on roads. Certainly roads come before laws, but they are useless and even dangerous without the white or yellow line – still a question of code – painted down its middle. The line is just an element of the road. But it is a significant element and probably the most significant one if we consider the functionality of roads.

The law plays the same function in society as lines do on roads: they participate in the social landscape and organise it. Good laws should be as visible as a good line is. Good laws draw the ways and leave to our discretion the choice of following one way or the other, but also leave open the possibility of making the wrong choice. This reconstruction of nature by law does not mean that nature has no influence on the law. When the road curves and when it becomes dangerous for one car to overtake another, the line is a solid line to prohibit such behaviour. But such a reciprocal influence has some limits in a regulated society: a road, at least a non-private road, should always have a line painted down the middle.

It has been observed that 'nature has only recently become a central concern in the discipline of sociology'.[1] By contrast, it can be asserted that nature and law have long had a common concern and reciprocal influence. Nature – I mean what is given to us before any interference of human activity – has influenced the concept of law itself. For some lawyers, it is even possible to derive rationality from nature as it stands, i.e. fundamental rules which are the source of all existing and permanent laws. This is the theory of *jus naturalis*. For others, mainly what we call the 'Historic School', such rules are nonsense but 'the law can always cure the reality under its yoke' as the eighteenth-century German lawyer Savigny said. In both cases, law appears to offer a cultural construction of nature.

Although this cultural construction varies according to the legal theory chosen, it implies a kind of legal imperialism. Take the example of 'non-law' which is defined – in comparison with law – as the absence of law in some human relationships for which the law would have been theoretically necessary. Examples of non-law (it is not permissible to evict squatters in winter in some European countries, non-marital cohabitation) are also good examples of legal imperialism: if squatters cannot be evicted it is because a rule made to protect the vulnerable temporarily prevails over a rule protecting the owner of a property. Regarding non-married couples, although their legal status is not the same as that of married couples, some such status exists and these people cannot be regarded as outlaws.

After these preliminary remarks let us come back to the issue of a legal approach to biotechnological innovation. The methodological *a priori* is to recognise that there is no reason why legal concepts, the legal systematisation of nature, should not apply to the societal issues raised by the development of biotechnologies as they have applied in the past to other economic or social changes.

What are in fact the so-called new issues raised by biotechnologies? To those who would say that such technologies have opened the door to artificial reproduction, surrogacy or the possibility of 'manipulating' the human being, it can be objected that culture and imagination from the Bible to Aldous. Huxley have already posed such issues before the technologies to effect them were created. Of course, what is at stake in all these examples is an ethical value, a reference to the ontology of the human being – and not only 'practical' issues such as the degree of safety associated with an innovation. New problems will always appear as science evolves and it is necessary that legal rules take this evolution into account.

But when ethical issues are at stake the law takes on an ethical meaning. It formulates society's view not on the technology itself, but on what such technology represents symbolically. The way in which artificial reproductive technologies are approached today depends on the strength of cultural attachment to social or biotechnological parenthood. To change the law implies that society has developed a strong and real will to do so. A common mistake would, however, be to see such signs of the need for change merely in the evolution of social habits or in a strong political event such as a political vote. Changing the law needs other criteria, the most important of them being that the existing law cannot be interpreted to cover the qualitatively new issues posed by technological innovation.

Legal conservatism, as I will outline below, can then mean flexibility and creativity. However, very often, this flexibility, because it uses technical concepts and mechanisms, is not readily intelligible to the general public. Consequently, safeguarding the common good, which is a recurrent issue for the law, is assumed to depend on the quality of public deliberation. My interest below is in exploring how these two aspects articulate in the present debate on biotechnology issues and in the challenge raised to legal concepts by the multiple uses of the human body and its components for biomedical and biotechnological purposes.

THE CONSERVATISM OF THE LAW, THE INNOVATION OF BIOTECHNOLOGY

Beer is the outcome of a biotechnological process but also of an historically long industrial tradition which dates back to the ancient Egyptians. It is therefore deeply rooted in our culture. What, then, is biotechnological innovation today and how does it affect our culture and our social organisation? To clarify this a necessary prologue is to evaluating the depth and complexity of the relationship between biotechnology and law.

Defining and Perceiving Innovation

Genetic engineering techniques involve the production of an indefinite number of DNA associations which are not fundamentally different from what could result from normal evolution, but which can be rapidly brought about due to a new capacity for the technological processing of components of living nature. Gene transfer, and the expansion of genetic reconstruction which results from it, has opened the way to considerable innovation in agriculture and nutrition, in biodyollution, in the pharmaceutical industry and also in human therapy, making possible gene therapy and providing a better knowledge of human genes and defects through the project of mapping and sequencing the human genome.

This mastery – although still limited – of life and human life has raised not only questions and debates but also fear and anxiety in the public with regard to what is often called 'genetic engineering'. This is obviously not the first time in the history of the relationship between science and society that scientific applications have been considered risky or challenging to social values. But there is a great difference between what can be called the Galileo approach and the present approach. In the first case, Galileo's discovery and theory challenged the view that the Catholic Church, which was the dominant authority and ideology, had on the organisation of the universe. In the second case, the reactions to biotechnology in the West come from what we now call the public and from amongst the public, various groups of citizens. It proves that, although our interest in biotechnology is inspired by special insights, the awareness of their potential applications is greater, because people are now able to imagine more easily the potential side-effects on health, food and the environment than they could have imagined the impact of Galileo's theory on their everyday life. Another reason is that public opinion and citizenship today make a great difference to the manner in which science and societal issues are discussed. Consequently, what is at stake is no longer the institutionalisation of a hegemonic system of social and ethical values, since it is the right of individuals or groups to press for their own views to be taken into consideration in biotechnological applications.

From this perspective, from the pre-legislative step to judicial control, the legal process is often viewed as important because it offers procedural means to the citizen to enforce his/her rights and to impose some control on science and technology. Consequently, we are far from what is really the conservatism of the law.

An Operating Concept of Legal Conservatism

A French professor of law once wrote that the best thing he could ever teach lawyers is that they could never be as good as their predecessors. This opinion should not be misunderstood: it is not evidence of magisterial arrogance, but is simply the view that it is not the law – its concepts and the way to apply them – which has to be changed in the face of any kind of innovation. The lawyers and, via their mediation, society, can find in existing legal theory the means to approach and regulate what appears to be new. Consequently, the application of the theory of legal conservatism does not imply that we should dispense with innovation. It is not an anti-modernist approach with ethical implications such as the condemnation of gas lighting by the pope in the nineteenth century.

This conservatism is even beneficial to the social acceptance of new technologies because it offers a ready-made approach to any kind of innovation. Let us take the example of the car. If the risk created by this new engine could be in some way acceptable for drivers, at least when they were only a few 'crazy people' – the risks for non-drivers served to clarify who should be responsible in case of an accident. In adapting the 1804 French civil code provision on civil liability to this new technology, the French legal system, using general principles of law, contributed in its own way to the development of the car industry. In the biotechnological field itself, an example of the flexibility of legal conservatism is the decision of 3 October 1990 of the Technical Appeal Board of the European Patent Office in relation to the claim of 'Myc Mouse', a genetically modified mouse. The Board decided that the Examining Division was wrong to refuse the patent on the basis of Act 53b of the European Patent Office which excludes the possibility of patenting 'animal variety'. The Board considered that this exception should be strictly interpreted, balancing on the one hand the arguments relating to the suffering *created* for animals and potential damage to the environment with, on the other hand, the benefit for humanity of such an invention. It took two years for the Opposition Division to agree to grant the patent, acknowledging that the balance lay in favour of the invention, which was aimed at contributing to the development of new treatments in the field of human oncology. (Official Journal, European Patent Office, No. 10, 1992, p. 588).

Methodologically, the experience of legal conservatism is also very interesting since insofar as it uses substantial legal principles to approach an issue, it also uses a large range of legal mechanisms to permit social discussion and to reach a legal solution that can be enforced, through a system of legal sanctions. Contractual arrangements and court deliberations play a large role in defining the legal nature of new technologies and the rights and obligations that can be derived from their introduction in the legal judicial world.

However, the rationality of legal conservatism has its limits. The first is a time limit. The legal conservatism approach assumes that for those people who are not convinced *a priori* of its superiority, they will nevertheless be convinced as time passes by the experience of the method which will reveal its efficiency. However, in our world – where we expect everything to come to each of us instantly without any mediation except from what we call the mass media – it has become inappropriate to admit the conceptual necessity for the law to refrain from giving immediate solutions to what is presented as a new problem.

The second limit is a space limit. Our ancestors in law, the Romans, long divided the world into two: the Roman world with law as a founding value and the Barbarians with nothing. Although the border of the Roman world could be extended, our world has no closed borders, but constructs permeable borders from geographical and cultural components.

Therefore, legal conservatism as a methodology has to address what may be called internet symptoms. As it is now possible to have easy access to any type of information produced anywhere in the world which answers any specific issue as soon as such an issue emerges, the question is, why should we use an old-fashioned methodology supporting the application of concepts which appear related to a specific and limited social content? In other words, the complexification of the legal context – especially in the law-technological field – makes less efficient, and even less productive the legal conservatism approach, which is based on a continuous search for consistency between pieces of legislation which have not been drafted according to this perspective. For example, in the field of intellectual property, as the US approach is more flexible on the concept of utility – what the Europeans call the industrial application – of the invention, the scope of what could be covered by a patent could be quite different in the US and in Europe. This comparison illustrates what may be called the deconstruction of the legal system which affects the principle of legal conservatism.

The Deconstruction of the Legal System

The deconstruction of the legal system is the result of a combination of four phenomena. These are:

1. The diversity of the interests that should be individually taken into account, which has never been so broad. The interests at stake are no longer simply those of direct influence to individuals, i.e. those legally protected in relation to the relevant issue. Any kind of social interest for which individuals or groups (I should say, lobbies) would demonstrate in the public arena should also be considered. As a consequence, the distinction between individual interest and the 'general common good' has lost its primary meaning. Groups of citizens often claim they are acting in the name of such a 'common good' while in fact adopting a *'laissez-faire'* policy. Public authorities are accused of allowing the standpoint that 'what is good for General Motors is good for the country' to prevail. New interests have also appeared: interests defend-

ing the rights of animals, interests dedicated to protecting nature, interests defending the integrity of the human species.

2. The legal doctrine – those lawyers influential in initiating new legislation – has lost its moral authority and has been substituted by the concept of expertise which in practice means a huge number of professional experts, each dealing with a specific field, demanding new and specific regulation. Although we need expertise, especially in technologies, to help us to understand how such technologies work and could affect our lives, credibility granted to expertise should be limited to what is the necessary mandate of an expert; to bring knowledge to a given situation and to influence opinion. By contrast, law is more than just a technique. It is a tool for organising social activity as a whole, using concepts and values which are not purely determined by the necessities of legal technique itself, but which are rooted in fundamental values.

3. Following the same logic, the legislator cannot now act in splendid isolation, but the legislation or the regulatory process must take account of the views of a variety of national and supranational bodies, and refer to binding rules rather than suggested guidelines, to fundamental principles as distinct from technical regulations, to public provisions rather than private ones.

4. There is great diversity in how and how far regulatory rules are enforced. There could be indirect sanctions in that a grant could be re-funded or withdrawn; there could be different legal sanctions, administrative, civil, criminal; there could be different jurisdictions competent to enforce the rules, whether domestic, international, private or public. All these mechanisms could raise potential conflicts – what international lawyers call forum shopping.

Let us now return briefly to the heated European debate about the patenting of biotechnology to examine how it fits in with the above analysis (see McNally and Wheale, this volume). With respect to diversity of concerned interests, it can clearly be affirmed that the rejection of the initial draft Directive by the European Parliament in 1995 was largely due to the fact that the variety of interests was not taken into account on an equal footing. The initial text was largely inspired by the desire to facilitate the protection of biotechnological inventions, while the reaction of the European Parliament was inspired by political opponents with the idea of protecting free access to nature and freedom of research. The second draft presented in December 1995 by the Commission was an attempt to achieve a better balance between the different interests concerned, in particular by clarifying the old confusion about what is a discovery and what is an invention in the field of patent law.

This clarification was also needed because some of the experts in patent law who drafted the initial text failed to write down the legal conditions under which the patenting of living organisms could be permissible. This reference would have certainly prevented the confusion generated by those who raised the legitimacy of such questions. Another difficult aspect of the EC Directive is that it ignores the existence of other European legislatures, essentially the European Patent Convention, which addresses the question of patenting biotechnology and has judicial mechanisms developed to solve it. To reduce this question of conflicting bodies of legislation, the second draft was compiled in collaboration with the European Patent Office. The issue of the enforcement of the specific rules adapted to patenting biotechnology is not yet resolved because it will rely on national legislatures and could also require amendments to the European Patent Convention. However, greater co-operation between all the parties concerned will facilitate this step. If this were to happen, it might be possible to limit the deconstruction of the legal system at the European level or to contribute in some way to its new construction.

The construction or reconstruction of the social reality which is produced by law is not superior to other types of construction using different approaches. Law may be regarded as an efficient – or at least less inefficient – way in which to organise human activity on clear and operating grounds and mechanisms. The problem is that the regulatory process has lost a large part of the role of clarification through the medium of law that it hitherto possessed.

EUROPEAN PUBLIC POLICY, BIOTECHNOLOGICAL INNOVATION AND THE SUBSTITUTION OF FORMAL LEGITIMACY

Given the situation that obtains in the European Union of a diversity of countries with different substantive laws, how can one single law on a very sensitive issue be produced upon which there would be some sort of consensus? It is generally the kind of provocative question – it's better to say assertion – given by those who feel defeated by – in those cases in which they are not explicity opposed to – possible attempts to harmonise European laws. As an alternative, they often stress the necessity to develop a more democratic legislative process, while they abandon the idea of promoting clear common values which will be the founding pillars of substantive European rules.

Formal Legitimacy: a Motto for European Public Policy?

The issue of legitimacy can seem ambiguous in relation to European institutions. There are no legal constitutional arguments to challenge the legitimacy of European institutions which would be substantively different from that used to criticise the national representative systems. It has to be kept in mind that while the concepts of sovereignty and representation are partly fictions, they are useful fictions. If we argue that they are only fictions, there is a risk of destroying the whole system. And, as Churchill said, democracy is a poor system, but there is no better one to replace it. Sociologically, the question of legitimacy can be problematic, especially in a field where decisions more often rely on personal choices than on a political party programmme.

Legitimacy, whether at the European level or otherwise, raises isues of participation. If it seems obvious that professional communities involved in the development of biotechnological innovation should not decide alone on issues which have profound significance for individuals and social organisation, it is not easy to define who should participate in the decision-making process and what this process should constitute. Should those who are supposed to benefit from the biotechnological applications participate? Who are they: health care 'consumers', consumers of the new food industry, farmers, industrialists? Is it possible to identify all of them? Are they organised? Can they be organised and can they support common interests, common values and common positions? But also, can those groups (or those who represent them) be considered as expressing views which reflect a widespread social interest?

The answers are not obvious. Although the increasing role played by the European Union is an incentive for interest groups to organise themselves as lobbies informing and pressing the European institutions in charge of policy-making, this trend has some limits, especially the limited jurisdiction of the European Union which is mainly concerned with economic affairs and the preservation of national traditions. We know that the political history of some countries – France, for example – does not facilitate the emergence of citizens' groups playing an active part in the process of defining domestic policies. But this does not mean that in the same countries public authorities do not have to consider the social importance and visible influence of some professional groups: farmers and the medical profession, for example. Therefore, those who think that they are not heard by the government often express the opinion that politicians only consider those with power and influence or those who are likely to demonstrate violently. This reveals a great paradox: the new information technologies are extending rapidly, offering access to unlimited information for everyone while at the same time societal communication has become a difficult task which could lead to political conflicts or create the impression that politicians are out of touch with the real world. It could lead to the political exclusion of some social groups. All this can easily justify our questioning of possible new ways to make political decisions.

All of this raises the question of whether a new or reframed decision-making process is possible that would be more relevant to the complex functioning of democracy. In my view, concerned groups and individuals should not have the same influence at all levels of the public debate, but such a debate should be structured to give a definite role to each actor. We can logically think of the steps of the debate including answers, assessment and decision-making. Envisaging a three-step process, professionals and lobbies would confirm their initiating role in the first step while they would be first-or second-class actors in the second step, depending on the role public authorities would like to play in it. Finally, their role in the third step would largely depend on the type of rules which are supposed to result from this process. Obviously the most socially recognised rules are adopted by the legislators while regulations are often the task of governmental agencies and guidelines a product of professional and academic circles. In the real world, things are in fact more confused and make difficult any attempt at rationalising and adapting the process to new 'democratic ways' because the strategies of the different actors concerned are very often unclear, while they are not reluctant at all to press such strategies.

Applying these insights to the issue of patenting biotechnological inventions at the European level, the influence of lobbies was great in the first and second steps in relation to issues of industrial concern, but issues relevant to citizens' groups became more prominent in the second and third steps. We can deduce from this that the failure of the entire regulatory process was probably due to the limited presence of lobbies in the first step and the exclusive presence of opposed lobbies in the second and third steps.

THE QUESTION OF ILLUSION AND THE NEW DEMOCRATIC PROCESS

If decisions are to be made by communities, should these be public authorities since there is a public interest involved? Which process of decision-making should be applied? Should it be the general process of public policy-making which implies that the administration and Parliament have a central role? Or should we consider that it is not possible to decide on these very technical issues in this 'traditional' way and consequently that there is a need for more specific bodies to deal with them?

Public authorities confronted with decision-making in biotechnology are usually not very keen to tackle these issues. They see the immediate difficulties raised by it with no hope of short-term benefit. Up to the present the pattern has been that if they are not obliged by the actions of diverse lobbies to reach a decision, they will not do so. The more intrepid will favour social discussion by setting up *ad hoc* or standing ethical bodies, while the less courageous will wait and see what interest groups will ask for action. But this political attitude can find some justification in the difficulties of applying traditional decision-making processes to a set of problems which cannot simply be answered by a yes or no. For example, what does the rule of the majority mean when Parliament is faced with the task of deciding upon fundamental values? Is it possible that a Parliament should define what life and death is? Is the referendum, which is supposed to be the most direct means to involve citizens in social decisions, an appropriate choice? The 1991 Swiss 'rotation' on bioethics and the 1998 one on genetics should be good examples to draw upon for reflecting on these questions.

The problem is that very often public authorities do not take advantage of this wait-and-see approach to elaborate efficient strategies to take account of future changes. Regarding the question of whether and how far it would be possible to patent biotechnology, it is true that the European Union included this issue in its programme at the end of the 1980s, but its strategy was purely economic and did not attempt to evaluate the ethical aspects or acknowledge any kind of public debate. Of course, it can be argued that it was not possible for the Commission to think that a technical question on the conditions required to get a patent fulfilled would have raised such a furore in the media. This argument is only partly true because experiences in similar fields such as the release of genetically modified organisms into the environment give evidence of a political context which is very sensitive to those types of issue, both in European society at large and in the European Parliament. And we can also note that when the biotechnological issue came to prominence in the USA, assessment of the related legal and ethical aspects was also made, in particular through reports of the Office of Technology Assessment to the Congress.

Are There New Ways to Conduct Public Discussion?

If we stick to the fact that in western, pluralist democracies, where it is not possible to found public policies on fundamental common values but it is necessary to accept negative consensus, the fight for a global democratic process appears to be a nonsense. But should we stick to such a limited perspective? I strongly believe that the demand for greater public participation in the decision-making process is a biased one and would have negative side-effects in so far as it has no clear political objective. For example, if the construction of a European institutional presence is an objective, it could mean that attempts at the national level to protect the domestic economy, although legitimate, should no longer prevail, even if the domestic debate might be more democratic or open to public participation than the European debate. Therefore, in the EU attempt to draft a Directive on patenting biotechnology, it should not be forgotten that the objective of this regulation is to benefit the entire Union and not to satisfy purely national interests such as the Greens or religous groups in Germany or in Scandinavia, farmers in France, Italy or Spain, or bio-industries in the UK.

It is, of course, possible to question the interest – international industrial competition – which the European Commission considered to prevail, but challenges cannot simply exclude such a content but will have to amend the proposal by balancing differently all the interests involved. The question is: how can a balanced policy be produced if each citizen group behaves as a lobby of individual interests and not as a proponent of the common good?

At the national level, in the absence of an agreed global objective that could guide the reframing of the democratic process, we could helpfully refer to the judiciary as a model, which offers the guarantee of due legal process. However, although the judiciary could be a useful influence in improving the decision-making process by using procedural rules, ensuring the existence and efficiency of a public debate, requesting submissions from all parties involved and offering them equal means to present their case, this suggestion should not be misunderstood. Policy-making is not an issue which should generally be dealt with by judges. The function of the courts is to resolve conflicting interests and the judicial processes should only support, at least in countries where justice is easily accessible and equitable, those who come with their cases to such a forum. But the judicial model has two main characteristics which makes it inappropriate for solving societal issues: it operates on a case-by-case basis and it only considers legal rules as a means of reaching solutions. Of course, there is something to learn from this process with respect to deducing from substantial legal provisions possible applications to new individual cases. The judicial process can only work in this way if the principles which serve as a basis for this reasoning are sufficiently clear and can be accepted as founding pillars for defining new framework policies. If it is not the case, it implies that society cannot find common values to achieve this through its legislative process.

At the European level, the task is even more difficult because the courts can only solve disputes on the basis of European legal rules. Consequently, the recourse to the judiciary would be of no help when what is at stake, as in the field of patenting biotechnology, is precisely the question of deciding what constitutes a common value basis for European law. The other way to explore the issue, one that is politically more efficient, would be to rely on the work of an independent body belonging to the category of ethics committees. Therefore the decision of EU President Jacques Delors to set up a group of advisors for the ethics of biotechnology in 1991 was certainly not a coincidence. This new group is an example of an innovation in the approach of the European Union.

At the end of the 1980s, working groups were set up in order to prepare guidelines for researchers claiming grants from the European Union in fields of research considered particularily sensitive: embryo research and research on the human genome. But those working groups had a limited mandate related to a specific research programme. They were also very dependent on the executive branch of the Commission in charge of scientific research for which they functioned as consultative bodies.

Where, then, is the space for citizens? In my view it is a mistake to believe the only space for citizens should be global political activism. Although the pressure put on Parliamentarians or on other European policy decision-makers is of some influence, as is shown by the drafting by the European Commission of the Directive on patenting biotechnology, this leaves unresolved the question of how to transform political will into a legal duty coherent with the wider legal system.

One way is already used by citizens' groups. This is the possibility they have in some jurisdictions, at least within the remit of the European Patent Office, to oppose a patent's claims. This possibility should probably be extended to other jurisdictions because it forces lobbyists to use legal arguments and consequently it could facilitate the production of a case-by-case regulation using the mechanisms of legal conservation. But this should not be misunderstood. All arguments should not be transformed into judicial ones and only discussed in a judicial process. The suggestion is oriented towards facilitating the use of exising legal frameworks before considering changing the rules. But other kinds of arguments, say political and economic ones, should also be heard in the earliest stages of the decision-making process. Those who, as institutional role-bearers, have the power to initiate policy should be constantly aware of the societal view of biotechnology. It is not clear that quantitatively-based Eurobarometer studies on biotechnological issues can really replace, for example, sociological research studies, with real interviews of the people concerned.

The above consideration is for the information-gathering and awareness-generating dimension of any policy. What, then, about the assessment dimension? There, too, more is needed not only to bring the stuff and the argument for any new policy to public attention, but also to offer the general public an independent assessment process in which they would participate and benefit from the output (data, research studies, etc.). In general, citizen groups have good views but bad arguments, compared to Eurocrats, who have good arguments but bad views. A corrective for this state of affairs is needed by offering citizens the possibility of benefiting from an independent (i.e. from the administration) assessment process.

It is interesting to note the attempt of the European Commission to operate with such a policy in the food safety area. For the first time in 1997, in recruiting the members of a follow-up committee, the Commission made a call for independent candidates, though they should possess an expert background to be considered for appointment by a panel working for the Commission.

While this is a progressive step, I am doubtful about the value of the amendment before the European Parliament regarding the European Directive on patenting biotechnology. The Parliament has suggested a standing committee to conduct an ethical review of all patents on living organisms. But once a decision has been made on the right policy to adopt – and it is the role of Parliament to do so – confusion should not be created: the case-by-case so-called 'ethical' review then becomes a legal review and should simply follow the rules *according to which* a patent is granted or not.

Searching for Common Values for Substantive European Rules

It is less the idea of common values in themselves that should be stressed as the fact that common values have to be associated with the perspective of bringing life to substantive European rules. As regards European common values, these values are clearly acknowledged in the founding texts of the European Institutions such as the European Convention on Human Rights, the Treaty of Rome and subsequent EU instruments. But, what is more difficult is to find in those foundational texts references which could introduce commonly accepted answers to each practical and sensitive issue to be faced at the European level. However, from my experience of participating in the work of international organisations I believe that such a task is not impossible if it is not intended to build a complete system of rationality on the spot. We should, therefore, accept giving life to harmonised regulations without explicitly deriving them from human rights values. It should also be admitted that some values are so distinctive that they could prevent common regulations from being established.

How is the harmonisation of rules possible without explicating common values? Simply because law-making is not just a matter of choosing values but one of promoting interests which in the main should be public interests. Therefore, we can understand that when new technologies are implemented, it becomes the responsibility of governments and European authorities to regulate them in so far as the absence of regulations could have unpredicted or regressive effects on their social applications. The practice of blood transfusion and organ transplants offers a good example. The development of these techniques led physicians to search for rare blood products or organs outside their own country. Making this possible implied that the available products or organs could be easily used. It meant that common medical and technical standards – for example, blood products should be labelled consistently – had to be defined and accepted by all parties interested in the distribution of the products. This task has been mainly assumed for 30 years by committees of experts within the Council of Europe.

The independent definition of common standards does not mean that such standards cannot be implicitly related to common values. However, it is not required here to determine what such values are and how they would apply. This flexibility is made possible by a variety of factors: the need for regulation is urgent, the scope of regulation is limited, this regulation is made by and for professionals (mainly physicians) who are supposed to have been schooled in professional ethics, and the responsibility of European policy in this field relies on the Council of Europe, an organisation known for its human rights approach. But, although year after year the ethical principles of blood transfusion and organ transplants merged – using anonymous unpaid donors, developing national self-sufficiency, non-profit organisation – the time came when it was necessary to assert clearly and strongly the principles which regulate the system, particularily when there was a possible conflict with another system of rules. This was the case when in the mid-1980s the European Commission decided to prepare a Directive with the aim of faciliating the circulation of products obtained from human blood and plasma. The text finally adopted on 14 June 1989 was then criticised because, although referring to the 'code of ethics' developed by the committee of experts of the Council of Europe, it considered those human-derived products as drugs that should circulate on the 'European Free Market'.

The second draft Directive of the European Union on 10 December 1995 on patenting biotechnology now refers explicitly to common values – human dignity, the principle that the human body is *res extra commercium* – with the expected advantage of communicating the following potential message to opponents of the bill: the practice of granting patents in Europe does respect the above-mentioned values because the regulations according to which patents are granted are themselves founded on those values. However it could be questioned whether this politically inspired strategy is not coming too late given the difficulties of countering irrational arguments used by some politicians.

When Distinctive Values Outlaw Common Regulations

As the encyclical *Evangelium Vitae* of March 1995 stressed, there are legal enactments which are illegitimate regarding the respect of fundamental human values such as the right to life. Although the encyclical speaks of a moral legitimacy, the European Commission was clearly aware of the risk that such a theory could have on its research policies in the biomedical sciences.

Paradoxically, while the Council of Europe drafted a European Convention on specific regulatory issues, the European Union (as mentioned above) chose to set up expert groups in the specific fields of embyro research and human genetics to favour the adoption of guidelines. The limited jurisdiction of the European Union in this area as well as its own preference for pragmatic rules which would not set binding limits to research are certainly an explanation for this 'strategy'. But this is also the consequence of the long-term absence of a strategic perspective regarding the political role of values – such as human rights – which were never perceived by EU institutions as having possible direct effects on research or industrial policies. The facts are that today embryo research has been recognised as a matter of domestic concern. It is largely banned in Europe by national legal codes, although it is legal under specific conditions in some European countries. In this example, as in others, the more the values seem fundamental, the more the last word seems to be given to the national jurisdiction. The same trend appears also in the drafting of the European Convention on Human Rights and Biomedicine which includes in some of its provisions an explicit reference to national regulations.

The Necessity of Regulating Biotechnology

The difficulty of regulating biotechnology compared to other scientific applications is that biotechnology is viewed as a Pandora's box. It allows not only interventions in nature to repair the damage caused by industrial activities, but it permits remodelling of humanity and its natural environment. That is why we fear human germ line gene therapy or the genetic manipulation of plants.

The question that brings law and any human activity, such as biotechnology, together has always a practical finality: it is to regulate this activity with the objective of giving it a social perspective. Biotechnology and its applications are not meaningful in themselves. They have a meaning in a determined social and cultural context: the industrial world and the consumerist society. Therefore, their development implies that we live in societies where human intervention in nature is accepted as a founding principle and that human freedom is regarded as a common rule. However, because this freedom implies that we shall use it in an impossible way, regulation is a necessity of social life. Regulating biotechnology is then a counterpart to prevent abuses that would affect human beings but also other living organisms, as well as the global environment and future generations.

CONCLUSION

In conclusion, following from the above discussion, I wish to focus on a number of considerations which pertain to what is a key and necessary innovation in the biotechnological field: adequate European regulation. In the first instance, clarity on the main reasons which indicate the necessity for such innovation is required. It does not derive from the necessity to foresee common technical standards or to legitimate what is not yet clearly permissible. Existing international and European intellectual property law as well as Europe's own regulatory mechanisms could provide, if not entirely at least partially, some of the required new rules. The real justification for a new visible European legislation is that biotechnology is a crucial area of competition between Europe and its challengers, the United States and Japan, on the international world market. The European industry therefore clearly needs an evident and vigorous incentive to face this challenge.

From an economic standpoint, European biotechnological regulation must happen. That is certain, but is there any other impulse – and especially as regards the EU policy approach – that could justify this compulsory move? The answer is much more uncertain because the economic trend which is the real incentive for the European regulatory process is provoking a deconstruction of the legal process as outlined earlier. In reaction to the pre-eminence of economic views stands the political activism and 'altruism' of some citizens' groups and their political representatives. But this, as the failure in 1995 of the European Parliament to approve the first proposed Directive on patenting biotechnology shows, is only confronting an economic pre-eminence with a 'political' one. It is not building a new legal framework directly but it may open a way to it.

In the end regulation will come because a regulation incorporated in a coherent European legal system is of vital importance for the social acceptance of Europe as a set of institutions able to govern the people living in the European jurisdictions. Therefore, two conditions must be satisfied. The first one is to use the concept of legal conservatism as much as possible because it contains indispensable experience for introducing coherence into law. The second is to open the mechanism which produces the law to some form of legal debate which incorporates political arguments and transforms them into legal matters in order to regenerate and substantiate the legal order.

10 Bio-patenting and Innovation: Nomads of the Present and a New Global Order

Ruth McNally and Peter Wheale

THE DYNAMISM OF MODERNITY

In *The Consequences of Modernity* (1990) Anthony Giddens characterises the present time as 'late modernity' or 'high modernity', rather than post-modernity. For Giddens, the essence of modernity is its dynamism – a dynamism of such pace and scope as to be discontinuous with traditional social orders. He argues that one characteristic of modernity is the emergence of four institutions: capitalism, industrialism, surveillance and military power. Figure 10.1 sets out these four basic institutional

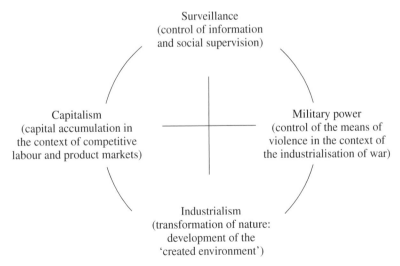

Figure 10.1   The institutional dimensions of modernity
Source: Giddens (1990, p. 59).

dimensions of modernity and indicates their interrelations. Innovations in science and technology generate utopian and dystopian expectations, which are sources of dynamism for each of these institutions. For industrialism, such innovations suggest new ways of controlling nature; for capitalism, new avenues of capitalist expansion; for surveillance, more penetrating technologies of observation and monitoring; and for military power, new ways of perpetrating violence. Elsewhere, we have applied Giddens's framework to analyse how utopian and dystopian expectations of genetic engineering are shaping modernity (see Wheale and McNally, 1994; McNally and Wheale, 1994).

In this chapter we focus on the dynamism in modernity which derives from the 'reflexive dialectic' – the clash of contradictions – over the benefits and risks of biotechnological innovation and patenting. In the first section we describe the perceived benefits of biotechnological innovation and how these benefits have influenced the interpretation of patentability to include genetically engineered life-forms. In the second section we describe a new global order which is emerging because of the patentability of genetically engineered life-forms. In the third section we argue that there is an imperative to patent which is not only creating a new global order but is unsustainable. Section four describes the emergence of a contemporary social movement opposed to the patenting of genetically engineered life-forms, which is challenging the hegemony of the bio-industrial complex.

'ANYTHING UNDER THE SUN'

Industrialism is the transformation of nature through the use of economic resources and mechanisation. In the early 1970s a new genetic engineering technique was developed for the transformation of nature by the use of natural phenomena. This 'microgenetic engineering' technique, which is called recombinant DNA technology, enables genetic engineers to 'read' and 'write' in the language of genes, and transfer genes between species (see Wheale and McNally, 1988a). Totally unrelated species that cannot interbreed can now be engineered with one another's genetic material. By comparison with traditional breeding practices, recombinant DNA technology is expected to reduce the time needed to produce new varieties of bacteria, plants, animals and viruses (see Wheale and McNally, 1986, 1988a, 1990, 1995; McNally, 1994, 1996). It is also expected to transform medical diagnostics and therapy through genetic analysis and intervention (see Wheale and McNally 1988b; McNally, 1995).

The novelty and perceived utility of the new microgenetic engineering is used to justify its use in basic research and for applications in agriculture, energy, chemicals, pharmaceuticals and the military, and is also playing a role in the industrialisation of health care and reproduction. This world-wide complex of scientific expertise, technological capability and transnational capital accumulation, operating on a global scale, constitutes what we call the 'bio-industrial complex', which is changing the experience of living in modernity (see Wheale and McNally, 1988a; 1994).

European Biotechnology Policy

Recombinant DNA technology is regarded as fundamental to biotechnological innovation, and biotechnological innovation has been identified as a new source of international competitiveness and economic growth in the European Union (EU). This was stated in the European Commission's (CEC) 1991 policy document on biotechnology – the Bangemann Communication – which states: 'Biotechnology is a key technology for the future competitive development of the Community and it will determine the extent to which a large number of industrial activities located within the Community will be leaders in the development of innovatory products and processes' (CEC, 1991, p. 1). Biotechnology was also identified (CEC, 1994) as one of three key technologies for European competitiveness, economic growth and employment.

Braun (1980) identifies a range of government policies for the stimulation of technological innovation. These include financial, fiscal, legal and regulatory, educational, procurement, information, public enterprise, public service, political, scientific and technical, and commercial measures. The policy recommendations of the Bangemann Communication (CEC, 1991) include many of these measures. (For a critical assessment, see Wheale and McNally, 1993.) Key recommendations include:

* public funding of biotechnology research and development, training and the information infrastructure;
* harmonisation of the legal and regulatory framework;
* the creation of a bioethics advisory committee;
* the creation of Community legislation on intellectual property rights.

Of these various measures, according to Maurice Lex of the European Commission, 'The harmonisation of patent protection in the Commission's proposal for a directive presents an essential element in the Community's multi-faceted strategies for biotechnology' (Lex, 1995, p. 235).

Innovation is believed to lead to economic growth, competitiveness and employment. Invention is believed to be an important prerequisite for innovation. However, invention requires investment in research and development, which is costly and risky. The patent system is believed to be an important stimulus to innovation because it is said to encourage investment in research and development leading to innovation by rewarding invention with a temporally limited monopolistic control. In the words an employee of SmithKline Beecham Pharmaceuticals, one of Europe's largest pharmaceutical companies:

> We would not be prepared to pour in the time and effort and the money if there was no possibility of securing adequate patent protection. (*File on 4*, Radio 4, December 1994)

It is also argued that the limited nature of the monopolistic rights and the disclosure requirements of the patent system stimulate diffusion. Consequently, absence of adequate patent protection is considered to be a disincentive for industry to invest in new technology, and a barrier to innovation (see, for example, Piatier, 1984).

Human Inventiveness

Biotechnological innovation is considered to be an important factor for future economic growth, and the patentability of biotechnological inventions is considered to be essential for the corporate investment in research and development that will lead to biotechnological innovation. However, at the advent of recombinant DNA technology it was not apparent that genetically engineered life-forms and their processes were patentable. In this section, we briefly describe recent interpretations of patent law so that living organisms – plants, animals and microbes – and their parts and processes, including the cells and genes of humans, are considered patentable.

The first step was the categorisation of industrially useful micro-organisms as 'inventions', which are patentable, rather than 'products of nature', which are not. The US Supreme Court deliberated upon this point in *Diamond* v. *Chakrabarty* 1980 and decided that the important distinction for patentability was not between living things and inanimate things but the distinction between products of nature, whether living or not, and human-made inventions. Thus, 'anything under the sun' was rendered patentable subject matter provided it was human invention.

The next step took place on the other side of the Atlantic, at the European Patent Office (EPO) in Munich, with the removal of the apparent exclusion provided by Article 53 of the European Patent Convention (1973) (EPC). (The EPC, which is a multinational treaty, is not part of the EU legislative framework.) Article 53 states that patents shall not be granted in respect of:

(a) inventions the publication or exploitation of which would be contrary to 'ordre public' or morality'; or

(b) plant or animal varieties or essentially biological processes for the production of plants or animals; this provision does not apply to microbiological processes or the products thereof.

Of crucial importance for the patentability of biotechnological inventions was whether the word 'variety' was to be interpreted narrowly or broadly. A narrow interpretation would be that a bar on varieties did not extend to larger categories, for example species, which contain varieties. The logic of the narrow interpretation is analogous to barring the patenting of fingers, but permitting the patenting of whole hands. Another ambiguity with regard to Article 53(b) is the interpretation of a 'biological process' (not patentable) as distinct from a 'microbiological process' (patentable).

The leading case in interpreting 'plant variety' is the 1983 Ciba-Geigy case in which the EPO's Technical Board of Appeal decided that the exclusion in Article 53(b) is to be interpreted narrowly. That is to say, if a type of plant is genetically altered, then it is not a variety, but a different plant altogether and should, for that reason, be patentable.

The next interpretive move was in the USA where the 'anything under the sun' principle was applied to animals. The first such patent was for a method for the production of sterile, polyploid oysters. This was followed by an eight-month moratorium on further animal patents in the USA. On the day the moratorium expired in April 1988, the Patent and Trademark Office (US PTO) issued the world's first patent for a higher life-form, a transgenic mouse known as the 'oncomouse' (see Wheale and McNally, 1990, 1995). Although referred to as the oncomouse patent, the patent claims all non-human 'onco-animals'.

Meanwhile, in 1988, a Committee of Experts on Biotechnological Inventions and Industrial Property established by the World Intellectual Property Organisation (WIPO), a specialised agency of the United Nations (UN), took the viewpoint that it was in favour of patent protection for biotechnological inventions, provided the usual conditions of patentability were met, i.e. novelty, utility (capability of industrial application), and non-obviousness (inventive step). That same year the CEC published the first draft of its Directive on the patenting of biotechnological inventions (CEC, 1988). This Directive, which incorporated many of the WIPO's recommendations (GRAIN, 1990, p. 7), was designed to establish unambiguously that living organisms are patentable inventions. Thus, Articles 3 and 4 (see below) of the Council of Ministers' (1994) Common Position on the draft Directive can be seen as attempts to remove uncertainty with regard to the interpretation of Article 53(b) of the EPC.

Article 3: Biological material, including plants and animals, as well as parts of plants and animals, except plant and animal varieties, shall be patentable.

Article 4: Uses of plant or animal varieties and processes for their production, other than essentially biological processes for the production of plants or animals, shall be patentable.

It was in the midst of uncertainty with regard to the interpretation of 'animal variety' in the EPC that in 1989 Harvard University filed an application for a patent on the oncomouse at the EPO. After being rejected by the Examining Division, going to the Technical Board of Appeal, and then going back to the Examining Division, the patent was granted in 1992. Thus, at this time the EPO appears to have interpreted Article 53(b) of the EPC to mean that genetically engineered plants and animals can be patented provided they are not varieties, and genetically engineered plants and animals were considered not to be varieties.

International recognition of the patentability of biotechnological inventions is also a feature of the UN Convention of Biological Diversity (1992), which came into force in December 1993. Article 16 of this Convention provides that access to and transfer of technology subject to patents and other intellectual property rights shall be provided on terms which recognise such rights. This Convention has been signed and ratified by 152 countries and the EU, but has not yet been ratified by the USA.

The patentability of biotechnological inventions was also included under the General Agreement on Tariffs and Trade (GATT). The GATT is an international agreement to liberalise trade in manufactured goods, services and agriculture. The Uruguay Round, concluded in December 1993, has been endorsed by 117 countries. The Trade Related Intellectual Property Rights (TRIPs) under the GATT extend western models of intellectual property rights to less developed countries. Developing countries will have a transition period of five or ten years in which to bring their law and administrative practices into line as a condition of membership of the World Trade Organisation. At this juncture we shall leave the story of the broadening of the scope of patent protection to consider some of its manifestations.

A NEW GLOBAL ORDER

By the early 1990s, the patentability of the products and processes of genetically engineered organisms had been established. However, there was
still one obstacle to the realisation of the goal of economic growth through
biotechnological innovation: genetic engineers do not create new life, they
re-engineer existing life-forms. In order to do this, they need genes and
living organisms – genetic diversity and biological diversity (biodiversity).
In this section we describe the global consequences of the pursuit of
genetic and biodiversity.

Genes R Us?

Genetic diversity is the variety of genes within a species, for example, the
genetic diversity of human beings exists inside human beings. The Human
Genome Diversity Project (HGDP) aims to collect DNA samples from
about 25 individuals from each of approximately 500 ethnic human populations (see Schomberg and Wheale, 1995; Jayaraman, 1996). Through the
collection of blood, tissue and hair samples, researchers have already gathered genetic data from a number of indigenous peoples including the San
peoples of South Africa, the Penans of Malaysia, Australian aborigines,
peoples of the Sahara, Latin American Indians and the Saamis of northern
Norway and Sweden (Vidal and Carvel, 1994). The Guaymi Indians from
Panama have been found to have an unusually high incidence of human
T-cell lymphotropic virus (HTLV) and cell lines grown from Guaymi
samples are considered to have a commercial use. In 1991 researchers
from the US Centers for Disease Control and Prevention (CDC) filed a
patent on a sample obtained from a 26-year-old Guaymi woman
(Anderson, 1993). In 1995 the US PTO granted a patent to the US
Department of Health and Human Services on a HTLV derived from the
Hagahai people of Madang Province in Papua New Guinea (Dickson,
1996).

The genomes of indigenous peoples are considered to be valuable
sources of human genetic diversity because of their relative genetic isolation. The assumption underpinning the HGDP is that genetic differences
between 'races' are more significant than those within them. The HGDP is
establishing a gene bank from endangered indigenous peoples before they
disappear in case they contain some genetic variant which could be of
scientific interest or utility in diagnostic tests and therapies. However, in
the words of Debra Harry, a Paiute North American Indian, genetic

research is 'not a priority for indigenous peoples. They've come to take our blood and tissues for their interests, not for ours' (Butler, 1995). The commercial value is not in conserving indigenous peoples *per se*, only their extracted genes. Indeed, once genetic samples have been taken, the peoples become devalued from the researchers' point of view and their genes can become the patentable subject matter of other agents.

A New Comparative Advantage?

Biodiversity is the range of different plant, animal and microbial species. It is the basis of almost all the world's food crops and drugs, and is the starting material for modern genetic engineering (see Juma, 1989; Fowler and Mooney, 1990). Ninety per cent of the earth's biodiversity is located in the less developed countries of the South – in Africa, Asia and South America. Less developed countries are 'gene-rich' but 'gene technology-poor', whilst advanced industrial countries are 'gene technology-rich' but 'gene-poor'. The biodiversity of less developed countries can be considered a new 'comparative advantage' which they could trade on the world market to advanced industrial countries (see Putterman, 1994). Advanced industrial countries would gain access to the biodiversity of less developed countries, whilst there would be transfer of gene technology from advanced industrial countries to less developed countries creating a global market in biodiversity and gene technology.

This principle appears to be upheld in the UN Convention on Biological Diversity 1992. The objectives of the Convention, set out in Article 1, are: 'The conservation of biological diversity, the sustainable use of its components and fair and equitable sharing of the benefits arising out of the utilisation of genetic resources, including appropriate access to genetic resources and by appropriate technology.'

Advanced industrial countries who transfer gene technology to less developed countries will have their interests – their 'comparative advantage' – protected through the internationalisation of the patent system under the GATT. However, the patent system does not protect the 'comparative advantage' of less developed countries. Although biodiversity is unique and is the product of the stewardship of indigenous peoples who often have recognised, protected, developed and utilised its potential, it cannot be left alone and patented *in situ* because the patent system does not reward conservation. Its uniqueness can only be recognised by the patent system when individual genes are identified, extracted, characterised and exploited through gene technology.

Whilst Article 15 of the UN Convention (1992) recognises the 'sovereign rights' of states over their natural resources and the authority of national governments to determine access to their genetic resources, there is no international system equivalent to patenting whereby less developed countries can protect their 'comparative advantage'. There is no system for the prevention of 'bio-piracy', which the Rural Advancement Foundation International (RAFI) (1994) estimates to be cheating developing countries of US$5.4 billion a year in royalty payments from food and drug companies which use their plant varieties and local knowledge.

The Biological Diversity Convention (1992) encourages bilateral contracts between developing countries and private companies. An example of one such agreement exists between Costa Rica and the USA-based company Merck Pharmaceuticals. In 1991 the National Biodiversity Institute of Costa Rica – INBio – signed a biodiversity prospecting contract with Merck. This contract gives Merck the rights to screen, develop and eventually patent new products from the resources (plants, microorganisms and animals) in Costa Rica's rain forests. In return, Merck has paid US$1.3 million to aid Costa Rica's conservation programme and has agreed to give INBio an undisclosed percentage of any royalties.

One view of such contracts is that they benefit both the advanced industrial partner and the less developed country and aid conservation as well. However, to put a perspective on the figures involved, Merck's annual sales in 1991 were US$8600 million (RAFI, 1994). Moreover, Costa Rica's stock of biodiversity is finite and cannot be recreated, and can only be sold once, whereas the number of innovations that can be derived from it is theoretically limitless, and can be sold or licensed many times to many buyers. Furthermore, if this trend were to be the model for the future, the world's stock of biodiversity could become the property of the handful of companies rich enough to purchase exclusive rights to it from less developed countries. Given that Costa Rica holds 5 per cent of the world's biodiversity, the entire global stock of biodiversity could be sold in similar deals for just US$26 million. Indeed, Shaman Pharmaceuticals has similar agreements with a dozen countries (Lemonick, 1995).

In what could be considered to be an historically epic moment (see Piore and Sabel, 1984), biotechnology has been identified as one of the key technologies for economic growth, employment and competitiveness in the twenty-first century. From the moment when biotechnological inventions came to be patentable, there were two types of life-forms: those which can be patented and those which cannot. The difference is that some life-forms are inventions, whilst others are not. What constitutes this difference in most cases is that genetically engineered life-forms are patentable, whilst non-genetically engineered ones are not. This legal distinction between the patentable and the unpatentable creates a hierarchy of life-forms with a premium on genetically engineered life-forms over others. Similarly, it creates a hierarchy among social actors, favouring those that have gene technology over those that do not. These hierarchies of life-forms and actors shape social relations. Biotechnological innovation constitutes a new 'regime of accumulation', and the globalisation of intellectual property rights in genetically engineered life-forms constitutes a new 'mode of regulation' (see Aglietta, 1979). Together they are creating a new global order.

THE IMPERATIVE OF PATENTING

The patentability of biotechnological inventions together with the discourse on biotechnological innovation and economic growth constitute an imperative to innovate and patent. This imperative is a new source of dynamism, and visions of possible futures resulting from it can be glimpsed by extrapolation from the present.

The American chemical company WR Grace & Co. has patented a method for extracting the active chemical compound azadirachtin from the neem tree, a tree which is used widely for traditional medicinal, contraceptive and pesticidal purposes by many indigenous communities in Asia and Africa. WR Grace and the US Department of Agriculture also have a patent from the EPO on a technique for using a fungicidal extract from the neem tree (Vidal and Carvel, 1994; Dickson and Jayaraman, 1995).

In October 1992, Agracetus, a company owned by WR Grace, was awarded a patent by the US PTO on cotton. 250 million people depend for all or part of their cash income on cotton production or processing (RAFI, 1994, p. x). Nonetheless, for the life of this patent, one company has monopoly rights in the USA over all transgenic cotton plants and seeds. In 1994 the same company was awarded a patent by the EPO on genetically engineered soybeans – a crop which was developed by Chinese farmers. Agracetus's patent covers all genetically engineered soybean plants and seeds and their natural offspring for the life of the patent which is 17 years. Patents on other major crops – rice, groundnut and maize – have also been applied for (RAFI, 1994, p. 13). These 'species patents' have been likened to Ford being given a patent on the automobile. Furthermore, there is concern that more crops will be monopolised by similar patents.

There are also attempts to obtain monopoly rights over human genes. One example concerns a genetic mutation associated with cystic fibrosis (CF). Through research on DNA from people with CF, it has been found that 70 per cent of them have a particular gene sequence which is referred to as the CF gene. Tests to identify carriers of the gene have been developed. A company formed by the Toronto Hospital for Sick Children has applied for a patent on the gene. In 1994 Manchester Regional Genetics Centre was sent a letter demanding a $5000 licence fee plus a royalty of $4 per test or 2 per cent of the fee charged to the patient (whichever is the greater) for every time the screening test using this gene is performed (*File on 4*, Radio 4, December 1994). Similarly, the University of Utah and Myriad Genetics have applied to the US PTO for a patent on BRCA1, a gene associated with breast cancer (Butler and Gershon, 1994).

Another example of 'genes as currency' concerns Human Genome Sciences Inc. (HGS) in the USA, whose Institute of Genome Research (TIGR) has compiled a databank containing 150 000 fragments of human DNA sequences – between one third and one half of all human genes. The main condition of access to the TIGR database is that the institution for which an investigator works must sign an 'option agreement' with HGS, under which HGS will have an exclusive option on any patents arising from research using the database sequences (Dickson, 1994).

Many scientists are opposed to the patenting of human genes but because human genes are patentable, and because others are patenting them, they feel obliged to do the same. If they do not, not only could they find themselves paying royalties to the patent holders, they may even find whole areas of research unavailable to them because of the cost of the licence fees. Thus, at the same time as patenting is a stimulus to corporate investment in research and development, it is also a stimulus to patenting itself, even amongst those opposed to the principle of patenting, because if one organisation does not patent a process or product, another one might. As Professor Martin Bobrow of Guy's Hospital complains: 'We, and many other University Departments and public research institutions are having to spend money taking out patents because we cannot afford not to' (*File on 4*, Radio 4, December 1994).

Sustainability?

If patenting begets patenting, how sustainable is this trend? Let us imagine each patent as capturing an inventive field, which then becomes monopolised for the life of the patent. Until a particular field has been patented, there is competition, investment in research and development, innovation and invention. Bit by bit, the inventive fields of biotechnology are becoming monopolised: Merck has a monopoly over 5 per cent of the world's biodiversity through its agreement with Costa Rica; Agracetus has a monopoly over genetically engineered soybean; and Mansanto has a monopoly on the use of neem as a fungicide and insecticide.

Each new patent monopolises a field of research and acts as a disincentive to corporate investment, a barrier to entry, in that field. This is because once a patent is granted, the incentive – monopolistic control – is removed until the end of the life of the patent. Furthermore, the licence fees and royalties exacted by patent holders are further inhibitors of research and development in the patented field. Innovation theory predicts that in the absence of the incentive of monopolistic control through patenting there will be a withdrawal of corporate investment in research and development resulting in the collapse of economic growth mechanisms. From this perspective, while patenting and genetic engineering together are, at the present time, contributing to a phase of rapid expansion in capitalism and industrialism, this trend is not sustainable for the reasons given above. However, just because a trend is not sustainable does not mean that it is does not have an effect. Together, genetic engineering and patenting are contributing to the dynamism of modernity. They are reshaping the face of capitalism and industrialism, and establishing a new global ordering of life-forms and agents.

NOMADS OF THE PRESENT

To return to Giddens's (1990) model (see Figure 10.1), driven by utopian expectations of the science and technology of genetic engineering, capitalism and industrialism are undergoing rapid expansion. Such transformation is underpinned by the expectation of social benefits – profitable products and processes, and unique ways of transforming nature to meet human needs and demands. However, as described above, this transformation also engenders a new array of high consequence risks. Giddens (1990), however, posits an alternative vision in which modern institutions undergo utopian, rather than dystopian, transformations.

The question then, is how to 'harness the juggernaut' of modernity, or at least direct it in such a way as to minimise the dangers and maximise the opportunities which modernity offers to us?' (Giddens, 1990, p. 151). What is needed, suggests Giddens (1990, p. 154), is the creation of models of 'utopian realism'. Utopian realism, he argues, could generate an alternative trajectory of social development which could lead to a utopian, rather than dystopian, future. The argument is that the alternative futures envisaged through utopian realism can transform the future because their very visualisation might help them to be realised. Key players in such a transformation are social movements. Whilst they are not the necessary or the only basis of change which might lead us towards a safer and more humane world, they are able to provide us with glimpses of possible futures and are in some part vehicles for their realisation through their reflexive dialectic with the modernation state and the scientific establishment (Giddens, 1990, p. 161; Wheale and McNally, 1994, 1996; McNally and Wheale, 1994). Contemporary social movements are described by Alberto Melucci (1989) as 'nomads of the present'. He characterises such movements as a kind of 'new media' which 'take the form of networks composed of a multiplicity of groups that are dispersed, fragmented and submerged in everyday life, and which act as cultural laboratories' (Melucci, 1989, p. 89). The networks are founded and live from day to day upon the daily production of alternative frameworks of meaning (Melucci, 1989, p. 70). Whilst their mere existence is a latent challenge to the ideology of the institutions of modernity, such movements also have the potential for resistance or opposition when a field of public conflict arises (Melucci, 1989, p. 70). At such times they emerge from their latent state and become visible as social organisations which present the rationalising apparatus of the institutions of modernity with questions which are 'not allowed'.

The emergent social re-ordering through biotechnology and the patenting of genetically engineered life-forms has engendered a social conflict which has provoked into action, and thereby rendered visible, a new social movement. This new social movement is nourished by a broad cultural network of values for reasons described by Genetic Resources Action International (GRAIN): 'Be it agriculture you are concerned about, or the environment, animals, the Third World, religion, or humanity in general, the patenting life debate has something relevant in it for you' (GRAIN, 1990, p. 17).

The social movement in opposition to the patenting of life-forms employs a diversity of strategies, as the examples below illustrate. Legal oppositions have been made. For example, in 1995, RAFI filed opposition at the EPO to WR Grace's soybean patent, and a request was made by more than 200 aid and environmentalist groups to the US PTO to withdraw WR Grace's patent on a method for extracting a neem tree pesticide (Dickson and Jayaraman, 1995). A further legal opposition to the EPO was filed by the Greens of the European Parliament, the International Federation of Organic Agriculture Movements, and the New Dehli's Research Foundation for Science, Technology and Natural Resource Policy against WR Grace's patent for a fungicide derived from the neem tree.

Petitions and declarations have been signed. These include the Joint Appeal Against Human and Animal Patenting, launched by leaders of 80 religious organisations, including Catholic, Jewish, Muslim, Hindu and Buddhist faiths in the USA (Emmott, 1996), the Farmers' Charter issued by Indian farmers at the Kisan Mahapanchayat, and a formal petition which a worldwide coalition of more than 250 women's health and social justice groups plans to file with the US PTO challenging the BRCA1 patent application.

Demonstrations, both in Europe, particularly at the EPO in Munich, and in other parts of the world, have taken place, frequently featuring animals represented by humans (see Figure 10.2). Spectacle, this time representing tradition and collectivity, was also a feature of a demonstration in 1993 when farmers throughout the Indian state of Karnataka gathered with drums and trumpets at the offices of each District Collector to demonstrate for the recognition of Collective Intellectual Property Rights, an alternative system which treats knowledge as a social, collective product and thereby constitute a challenge to the patent system. The subversion of technocracy by technology was employed in 1995 when scientists at Washington University, Missouri, USA, protested against gene patenting by putting a gene sequence involved in breast cancer susceptibility on the Internet (Dickson, 1995).

Information leaflets and booklets have been distributed. Figure 10.3 shows both sides of a postcard distributed by the British Union for the Abolition of Vivisection (BUAV) and Compassion in World Farming (CIWF) to encourage the public to register opposition to the patenting of animals by mailing the card to the EPO. On one side of the card is a computerised depiction of oncomouse, a 'cyborg' (Haraway, 1985) designed to die. The other side functions as a traditional lobbying card. Figure 10.4 shows two other lobby cards which symbolise the industrialisation and commodification of animals (although the oncomouse in both cases is actually represented by an 'oncorat'!). The strategic power of both the pictorial representations of technologised animals and the presence of dramatised animals at demonstrations is that they constitute a symbolic protest which cannot adequately be countered by legal documents or scientific reasoning.

These strategies used by the social movement have some successes in challenging the trend towards the patentability of life-forms. Notable outcomes of these strategies are described below.

Plant Patents at the EPO

In February 1995, the Technical Board of Appeals of the EPO ruled that a patent granted to Plant Genetic Systems and Biogen Inc. on a procedure

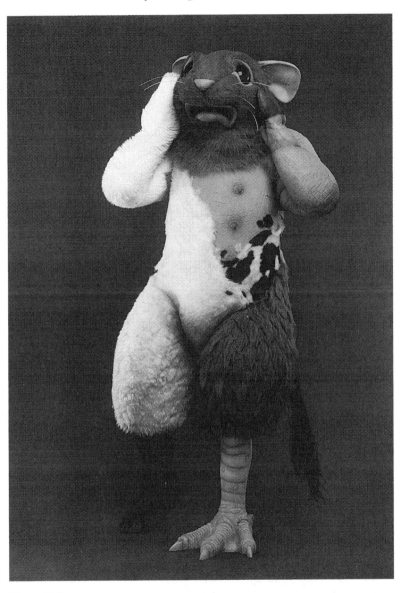

Figure 10.2
Source: British Union for the Abolition of Vivisection.

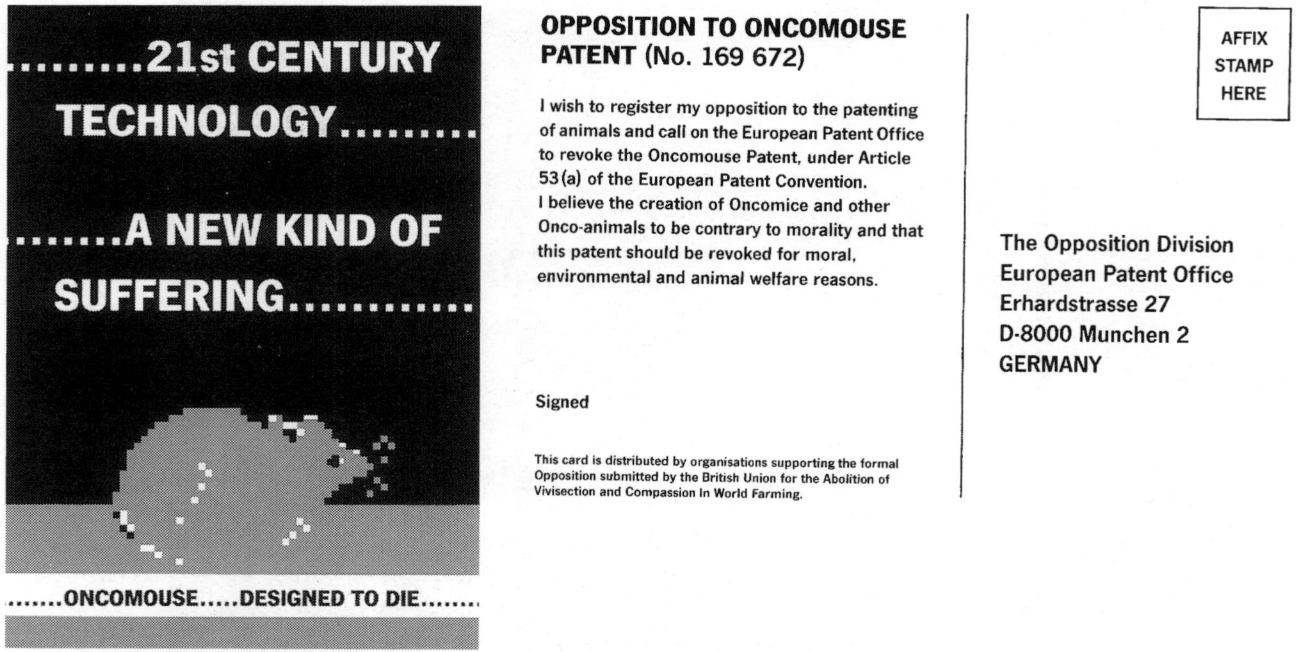

AFFIX
STAMP
HERE

The Opposition Division
European Patent Office
Erhardstrasse 27
D-8000 Munchen 2
GERMANY

OPPOSITION TO ONCOMOUSE PATENT (No. 169 672)

I wish to register my opposition to the patenting of animals and call on the European Patent Office to revoke the Oncomouse Patent, under Article 53 (a) of the European Patent Convention.

I believe the creation of Oncomice and other Onco-animals to be contrary to morality and that this patent should be revoked for moral, environmental and animal welfare reasons.

Signed

This card is distributed by organisations supporting the formal Opposition submitted by the British Union for the Abolition of Vivisection and Compassion in World Farming.

......21st CENTURY

TECHNOLOGY.......

.....A NEW KIND OF

SUFFERING.......

......ONCOMOUSE.....DESIGNED TO DIE.......

Figure 10.3   Opposition to Oncomouse Patent
Source: British Union for the Abolition of Vivisection and Compassion in World Farming

Figure 10.4   Help Stop Animal Patenting
Source: British Union for the Abolition of Vivisection and No Patents on Life.

for producing herbicide-resistant plants through genetic engineering cannot cover the plants and the seeds. The ruling, which was made in response to a challenge by Greenpeace in 1991, was based on interpretation of Article 53(b) of the EPC, which excludes plant varieties from patentability (see above). This is the first time that the EPC has been interpreted broadly as meaning that plants themselves, as well as plant varieties, are not patentable under the EPC. The significance of this decision is that it could have a major impact on the scope of the 100 or so patents already granted by the EPO on other genetically engineered plants (Abbott, 1995a, 1996; but see also Roberts, 1996). Moreover, opponents of the patent on the Harvard oncomouse are arguing that the precedent of the broad interpretation of 'variety' should cover patented animals as well (Abbott, 1995b).

Animal Patents at the EPO

In 1993 17 formal legal oppositions to the oncomouse patent were filed at the EPO (see Stevenson, 1995). These oppositions, which represented more than 200 groups, mostly challenged the patentability of the oncomouse under Article 53(a) of the EPC – the morality clause. Opponents also claim that the patent is invalid because of its breadth, covering any mammal genetically engineered for increased susceptibility to cancer, including, for example, an 'oncogiraffe', for which there is insufficient disclosure, no moral justification, and no known industrial application. The opposition also claimed that the oncomouse should be seen as an animal species, and thus as a collection of animal varieties.

In November 1995, the EPO appeared to indicate that the challenge to the validity of the patent had some legitimacy. Thus, whilst no final decision has been given by the EPO, the social movement's challenge to the patent has had the effect of halting the granting of patents on genetically engineered animals by the EPO. Since the oncomouse patent application was filed with the EPO in 1985, the EPO has received more than 300 applications for patents on animals. Only three – including the oncomouse – have been granted. Decisions on the remainder are effectively blocked pending the outcome of the oncomouse case (Abbott, 1995b).

'Bio-piracy'

In 1994, the US PTO provisionally revoked Agracetus's patent on genetically engineered cotton as a result of an opposition led by the RAFI, and in the same year the Indian Government revoked Agracetus's application for a patent on genetically engineered cotton in India. In Ecuador in July 1996, the intellectual property rights treaty with the US was blocked when environmentalists, aided by deputies opposed to the accord, staged a peaceful 'sit-in' of the Congressional Chamber.

Human Genome Diversity Project

At the end of 1993, the US CDC abandoned their plans to patent a cell line from a Guaymi woman from Panama after a challenge by the RAFI (Anderson, 1993), and several Pacific nations are said to be contemplating asking the UN General Assembly to seek an advisory ruling on the morality of human gene patents from the International Court of Justice. The ultimate goal of such a move would be to persuade governments to introduce tighter restrictions on human genetic material in the intellectual property provisions of the GATT, which are due for review in 1999 (Dickson, 1996).

EU Draft Directive

On 1 March 1995, after actions by environmentalists, human rights groups, animal welfarists, women's groups, farmers, third world development agencies and other non-governmental organisations, the European Parliament exercised for the first time its new powers of co-decision-making under the Maastricht Treaty and voted against the EU draft Directive on the patenting of biotechnological inventions, thereby removing this particular Directive from the EU agenda (Emmott, 1996). Despite this defeat, however, the Directive re-emerged at the end of 1995 when the CEC published a new draft of it (CEC, 1995). However, as a result of the actions of the new social movement against the first draft, this new draft has been delivered into a politicised climate where the power relations it embodies have been rendered visible from the outset.

A new global order created through the patenting of genetically engineered life-forms is the source of public conflict which has provoked into action a contemporary social movement employing both traditional and symbolic strategies. The actions of this social movement have not only been successful in challenging particular patent laws and the legitimacy of individual patents, but function as a 'symbolic multiplier': their actions have made power visible. Just as the social conflict has unveiled the 'hidden face of power' (Lukes, 1974) – the latent networks of meaning – of the social movement, so the actions of the social movement have unveiled the 'hidden face of power' of the bio-industrial complex. By unmasking the 'hidden face of power' of the bio-industrial complex, the social movement has created the conditions of possibility for the renegotiation of the rules governing biodiversity, genetic diversity and genetic sovereignty.

THE JANUS FACE OF POWER

Biotechnological innovation is considered to be an important factor for economic growth, and the patentability of genetically engineered life-forms is considered to be essential for biotechnological innovation. This has precipitated a reinterpretation of patent law so that biotechnological innovations are considered to be patentable inventions. The result is a global gene hunt in pursuit of genetic and biological diversity, and the claiming of intellectual property rights over genes, organisms and species. Whilst, as we have argued above, this trend is not sustainable in the long term because of economic stagnation caused by the patent system itself, the unequal distribution of gene technology combined with the globalisation of patent law is establishing a hierarchy of life-forms and agents, resulting in a new global order dominated by the bio-industrial complex. This new regime of biotechnological patenting has generated a field of public conflict which has rendered visible a social movement opposed, for a variety of cultural, ethical and symbolic reasons, to the patenting of genetically engineered life-forms. Through its actions, this social movement has not only successfully challenged particular patent laws and patent claims, but has made the Janus face of power of the bio-industrial complex apparent, thereby creating the conditions of possibility for renegotiating the rules governing biodiversity, genetic diversity and genetic sovereignty.

Part IV
Concluding Reflections

11 Modernity's Organic Economy of Governmentality

Tracey Skillington

BIOTECHNOLOGY AND THE RECONSTRUCTION OF NATURE

The science of biotechnology today is a powerful force; it pervades every aspects of social life, either directly or indirectly. As an institutionally supported reality, it is hailed as one of the major animating forces of late modernity. The increasing invasion of science into food and flesh means that long-life tomatoes, bananas and super soybeans that resist germs, even if mutant, epitomise the food of tomorrow, while tobacco containing a gene for thinning blood and bananas equipped with a gene that manufactures a vaccine against hepatitis B represent the future face of medicine. We are glimpsing at a future where government and privately funded science have a growing capacity to produce cross-specific brain transplants from one species to another, make non-human animals produce human sperm through the transfer of sperm precursor cells from the large mammal into the testes of its smaller relative, prime pigs for routine use as donors for human organ transplants (xenotransplantation), goats for the medicines in their milk (molecular pharming), and build factories for the purpose of growing human skin. As newly discovered genetic altering procedures are proposed for commercial application, they are individually assessed for safety by such bodies as the Advisory Committee for Releases into the Environment, and the Advisory Committee on Novel Foods and Processes before authorisation is granted for commercial release. In such a context, we are told, there is no cause for concern. Any expression of public anxiety about novel developments is discounted by those who threaten us not to try to shackle science's autonomy or the 'human yearning to find things out' on the grounds that 'research that must go on'.[1] Frequently aroused to a passionate defence of esotericism, scientists find themselves actively having to defend the continued separation of expert systems from the everyday life they can so dramatically alter, and

187

by extension members of the public at large, the latter of which Habermas (1974, p. 282) once described as the 'inmates of closed institutions'.

This kind of professionally shrouded threat, however, cannot quash public opinion's current preoccupation with uncertainty – uncertainty about the consequences of human 'progress', of genetic engineering, and of instrumentalised reasoning more generally. The growing inability of expert systems to anticipate future conditions has undermined the power of science's 'excellence deception' and replaced it with a now more widely diffused sense of its 'truth fallacy'. The rise in the general level of knowledge available to, and consumed by, growing segments of the population at large has publicly exacerbated the fallibility of modern social structures. Science, in particular, has come under severe attack, even though it is not strictly the paradigm of science that is subject to assault but its involvement with modernity's orientation to control (Giddens, 1994, p. 215). The constant outpouring of calamities associated with modern science has recently fortified this culture of uncertainty and unavoidably added strength to millennial angst. Flaws in the purity of the antibiotic society leading to the emergence of deadly strains of pneumonia and tuberculosis, the outbreak of BSE (Bovine Spongiform Encephalopathy), E-coli food poisoning, and bizarre environmental ill-nesses including 'Multiple Chemical Sensitivity' pose huge moral quandaries for those of us who have to live with science's consequences into the twenty-first century. From the start, these catasphropes have leaned more heavily on the technicalities of science, politics and market economics, than on those of consumer protection or animal welfare.

The social relevance of science's expert knowledge still depends more on its ability to complement the economic interests and cultural models of power elites than on its legitimacy as a new corpus of knowledge to a sceptical public (see Barnes, this volume). Biotechnology's cognitive distance from everyday cultures testifies to the manner in which power structures, in addition to social, cultural and economic 'needs', act as se-lective mechanisms for internally generated alternatives in modern science (Bohme, van Den Daele and Krohn, 1978; O'Mahony, 1991). The recent development of Dolly the cloned sheep at the Roslin Institute and the biotechnology company PPL Therapeutics is testimony to the external determination of biotechnology. Inspired less by intellectual marvel than by the enormous financial stakes involved in being able to perfect the genetic copier and transform the biological foundations of the agricultural industry through standardisation in the modern mass society, this creature is a hallmark of the type of organic commodification to come. Such a trend is being enhanced by government's threat to withdraw funding for new research, forcing scientists to seek out alternative private funding for their work.[2] The current scramble by genetic scientists to establish limited monopolistic means of protecting their discoveries with legal patents with a view to strategic economic gain is a more general indication of less virtuous motives than the quest for new knowledge on the mechanics of life (see McNally and Wheale, this volume). In a climate where scientific research is progressively more embedded in the oscillations of the capitalist market, controversy is very likely to be its long-term suitor.

The response of hard-core scientists to growing complications and the intensification of risk has been that science cannot be blamed for the demand for and commercial application of its innovations. From this perspective, the discovery of nuclear fusion, on which the atomic bomb was based, was 'neither bad nor good. It was simply the discovery of new knowledge'.[3] Wherever one wishes to lay blame for the destruction of our 'edible heritage', the downside of scientific 'advances in knowledge' appearing a generation after their much hyped public début has sharpened feelings of insecurity now that science, the high priestess of industrial modernity, from the public's perspective, can no longer be trusted. There is a growing perception of science as having unlawfully eclipsed social and moral codes. Consequently, the current rejection of human cloning by leading biotechnological scientists on the grounds that it is 'fanciful and serves no immediate needs, does not inspire a great deal of public confidence.[4] For biotechnology's milieux of economic and social innovation, social needs for genetically altered produce and processes are as manipulable as the biological blueprints they intend to alter. The claimed necessity of genetically modified food has been defended on such exaggerated grounds as combating world poverty, minimising agriculturally-induced environmental pollution, and promoting sustainable development on the macro-level. But social needs are also manufactured indirectly, the product of neither propaganda nor the unabridged diffusion of scientific innovation, but that of a cultural complementariety that implicitly endorses the mutilation of nature. Implicit cultural and social pressures may arise that indirectly justify, for instance the discovery of the age gene, including the eventual absence of social welfare provision and state support for the needs of the elderly, the modern cult of narcissism and a growing disrespect for all things aged and decrepit.

There are ample reasons to pause and reflect on what are the long-term implications of such developments, or even more fundamentally, what kind of society is being engineered and routinely endorsed that supports the assimilation of the world of living even more firmly into the world of the artificial in the interests of the changing needs of mechanised production and culturally prescribed convenience? The neo-biological future poses huge dilemmas for the youth of today and the perpetually young of tomorrow, not least of which is the question of whether this is a world they wish to inhabit. What will be the consequences of extreme loss of control, now that the new era of bio-automation transfers the possibility of human error from the action stage to the conception stage and thereby greatly diminishes the time or scope for intervention and reversal? The effects on the natural environment already indicate restricted scope for action, as part of the nature's future bio-diversity is already been disposed of in the present. The future has to be more a question of collective responsibility (see Strydom, Delanty, this volume) than the hegemony of the economic neo-liberal individualism of the current era allows for, if nature as we know it is to survive.

THE 'BODY–MACHINE COMPLEX' OF TODAY

Particularly since the late nineteenth century, bio-politics has brought with it new discourses and practices of observing, manipulating, and regulating the interface between the body and the machine. There is an unprecendent coupling of once distant bodies and technology, as the ideology of technologistical determinism is reinforced by the politics of bio-technology and those discourse coalitions within the institutional contexts of decision-making and practice (see McDonnell, and O'Mahony and Skillington, this volume). From the nineteenth century onwards, there has been an improved tuning of the behaviour of both humans, animals and plant species to the rhythms of the machine as the 'destruction of nature' has become institutionally standardised and integrated into the workings of everyday life. The tyranny of mechanical order and technological acceleration now embrace most pockets of life, as biological patterns, as well as social ones, are restricted by the structuring effects of the locomotive of mechanical time (Nowotny, 1994). However, the bio-technological era brings with it ethical anguish amongst a general public concerned about the implications of this culture of extraordinary control. Biology and technology now coalesce to produce the surveillance, and increasingly intolerant society. Genetic screening for insurance purposes, genetic fingerprinting for criminal investigation, and surveillance technologies comprise a legacy inspired by a new unanimity of feelings of insecurity in the risk society (Beck, 1996). The plague of insecurity has led to a new genre of consumption, the consumption of protection and obsessive tendencies towards precautionary surveillance.

The supposed identification of the homosexual gene, the serial killer gene, and genes linked to obesity or old age, thereby warranting genetic tests and accompanying 'treatments' for unitary natural causes, add a whole new dimension to pre-social, naturalistic, or biologically based conceptions of the body.[5] The 'test society' (Stehr, 1994) seems to complement the current political mood, one increasingly bent on the discovery of uniformity and predictability in the age of risk, as revealed recently in the rhetoric of 'one global nation', the 'decent society' and the 'age of achievement' amongst political elites. Hence, the formal rationality of teleological action, that is, action steered towards instrumental achievement and success, finds an advantageous complementariety in the orientation complexes (O'Mahony, 1991) and knowledge structures of bio-technological science, conventional politics with its desire to reproduce the traditional structures of industrial modernity, and a more general expansionist economistic rationality. Their complementarity is more securely a cultural one than it is a political or even economic one, although this does not preclude the latter. Indeed, cultural complementarity has created the necessary minimum conditions of legitimacy for a milieu of economic and social innovation (Castells and Hall, 1994) for biotechnology that try to enhance the institutional opportunity structures for a continuing biological science-industry synergy. Governments, for instance, continue to finance biotechnological research perfecting the art of cloning and transgenics, and permit the sale of genetically modified food, despite enormous public opposition to these developments. As Haas (1990, p. 11) explains this phenomenon more generally, 'science becomes a component of politics because the scientific way of grasping reality is used to define the interests that political actors articulate and defend'. Hence, in terms of an institutionally significant cultural endorsement of the interfusion of biotechnology, politics, and the economic, the biotechnological 'revolution', with the continuing aid of economic and politic support, has the potential to become a new means of public administration, an organic economy of governmentality that does not depend upon the threat of direct force but on the politics of the gene. The latter holds out the possibility of elaborate control over the body social and regular intervention into everyday life practices.

One could postulate that the cultural complementariety between a variety of political, biotechnological, and economic actors is, to a considerable extent, informed by and oriented towards conceptions of the body and nature, that are articulated by the historically embedded tradition of positivistic science. Whilst the understanding of nature as subservient has remained intact, the early theoretical positioning of nature and machine as oppositional has been rendered passé by the new advancements in biotechnology. There is an ever-increasing complementarity and unparalleled intimacy governing the genetically based 'body–machine complex' of today (Seltzer, 1992). It is no longer sufficient for humanity to be organised according to the charter of technocratic discipline but agricultural animals must now also. To ever-increasing degrees, they are forced to combine and adjust to the violence of systematicity; of repeatedly organised and restricted movement, schedules and classification on the one hand, and the market principle of high productivity and efficiency on the other, as science marches on in its efforts to force disembodied nature, the machinery of production, and the body of consumer convenience to merge into the contemporary 'ideal'. The normalisation of the wingless chicken is soon to be enhanced by the concerted articulation of a commercial and scientific discourse that wishes to maximise the production of edible bodily parts. Agricultural cloning is set to further intensify the practice of standardisation in the name of 'sustainable agro-development'. Farm animals of the future are more likely to be of a 'designer' variety which have been assembled with the utmost potential for useability in mind. The conquests of science have reached the intersection of interior biological and exterior disciplinary systems as the genotype assemble line is launched, marking a new era in the political and science-based economy of reproduction, as they converge with capitalist expansion. The emerging composite body of rearranged parts, more at home in André Kertesz's playful photography, epitomises the high point of where genetic science, consumer culture, and contemporary conventional politics merge. Premature obituaries aside, the time structure of modern science's linear, homogenised, divisible continuum is being transferred, via the organic machine, from the realm of the biological to that of social organisation and control (Nowotny, 1994).

Science's age-old use of the mechanical metaphor to depict nature has been testimony to how politics and anthropocentric ideologies have for long pervaded this paradigm. Their presence have not only intensified the cultural association of nature with quiescent, objectified femininity (Jordanova, 1980, p. 45), but have allowed for the subversion of the biological needs of nature in the interests of specific cultural ones. The relevance criterion of scientific-industrial truth dictates contemporary society's perverse relationship to nature, its pleasure in exploding its prescribed limitations, in overcoming the biological as a source of identity, and in seeking out self-understanding instead in consumer representations of the body, in the artifactual (Seltzer, 1992). We are thoroughly instructed in our differences – class, age, intelligence, gender, sexual orientation, race and species – instead of being reminded of the body as our common affinity to nature. This might explain why the body is progressively viewed as sacrificial in the face of scientific techno-economic achievement rather than as sacred and the primary source of selfhood. This cultural logic sets the requirements of bio-technical and mechanical discipline up as amoral-neutral ones when, in fact, the latter are acquiring a distinct moral authority of their own over time, one that is not in any sense above moral positioning and relativist judgement.

BIOPOLITICS, SOCIAL LEARNING AND COMMUNICATIVE GOVERNANCE

There is currently a profusion of biotechnology-inspired and institutionally bounded collective learning occurring. Biotechnological innovation has given rise to new modes of understanding, and descriptions of the work-ings of biological life (cognitive learning), to revisions in our relationship to the normative order (normative learning), and to changes in the way we relate to reality (aesthetic learning).[6] Such processes of collective learning are both a perpetual source of change and a principle of social organisation and thus add to both the self-transforming and self-enhancing capacities of modernity. Whether, in fact, innovations in the field of biotechnology en-gender 'progressive' or 'regressive' learning processes for the institutional and indeed, for the pre-institutional context, has ignited an intense and ac-rimonious dispute in the public sphere. Many are disturbed by the fact that in the midst of the emphatically celebrated cult of speed and convenience, genetic engineering is likely to launch a new branch of commodity aes-thetics which hold out the prospects of an aesthetically 'perfect' world, understood in terms of modelability and virtuality rather than constructs of beauty (Welsch, 1996, p. 6). This new sugar-coated commodity aesthetic is thought to be a symptom of a broader truth, a changing consciousness of both social and biological reality's nature and its totality as a deeply indeterminate entity. From the standpoint of biotechnology's milieux of scientific, economic, political and socio-cultural innovation, the biological foundations of modern life are the most pliable of current raw materials.

Humanity progresses at a fast-forward rate in its defiance of the inherent destruction of mass production and the mechanisation of the body. Some have depicted this process, conversely, as an acute version of slow motion suicide and the current BSE crisis is portrayed as a clear symptom of such insanity. The speed at which apathy has replaced fear about the beef crisis in Britain has provoked alarm.[7] One could say that the greater threat to modernity today is not apocalyptic fear about the future of our food, our environment, and nature more generally, but complacency and the uncanny ability to turn a blind eye to global threats. The latter is not so much a psychological constant as an institutional reality. It is reinforced by the growing anonymity and volume of danger, the symbolic normalisation of nature's disfigurement and its embedded institutional endorsement. Government officials in both the Irish and UK Department of Agriculture, for instance, seem mainly ingrained with the mindset of technologically boosted production at all costs. Can one expect fairness and balanced reason to preside when these officials remain wedded to a concept of progress linked to biotechnological advancement, and patrons/nurturers of one of the more economically valued, yet environmentally destructive of industries? Crisis is unlikely to dissipate so long as the Department of Agriculture continues in its contradictory role as champion of the interests of food producers on the one hand, and the health of the public as consumers on the other.[8] Even in the information age, the consuming public possess few rights to be informed as to what they are buying, as the nutritional and safety aspects of buying genetically engineered potatoes, strawberries, tomatoes, and oilseed rape, or sugar beet, maize, and soybeans used as by-products in the production of other foodstuffs, are willingly forgone for the sake of shelf-life and commercial availability. Few are even aware of the fact that they are eating genetically modified foods, much less able to exercise a choice on whether to consume it or not. Instead, we depend on voluntary bodies of committed groups like Greenpeace and concerned individual experts who have learned to use science as a resource for political action and sound the alarm on dangerous food production processes and foodstuffs.[9]

Western bureaucracy transforms the peculiarities of guilt and responsibility for environmental atrocities into acquittal and 'unintended consequences' over the long term. A system of purposive rationality that actually expects environmental disaster is the complexion of our current system of administration. Destruction is in keeping with the institutional practice of the non-applicability of rules of allocation and the principle of a guilty party to nature's destruction. As with the BSE crisis, both the dangers posed by such catastrophes and blame for their outbreak are externalised by conventional politics, science, and the economy precisely at a time when the legal system is purportedly individualising responsibility for environmental pollution and danger (Beck, 1996). Public discussion in the British Isles about the BSE crisis was pervaded by an obsessive consideration of the possible collapse of the beef market and the mass slaughter of prized herds, matched only in messiness by the bloody-minded nationalism targeted at those who refuse to buy 'our' beef or dairy produce. British politicians, mass media and agricultural representatives alike lashed out at the 'green zealots' of Europe and 'irrational' and 'historical' European consumers supposedly trying to sabotage Britain's agricultural sector.[10] For the silent majority of institutional adherents, the primary function of this kind of economistic discourse and one of banal nationalism is reassurance and defence against deep-seated anxieties regarding the innate destructiveness of a social system that pits its survival mechanisms against that of inherited nature, despite a recent shift in rhetoric on the environment and claims to be committed to radical change on the part of the state.[11]

It is the trail of the side-effects of environmental disasters that lead many people to turn away from the conventional political system and its perpetuation of an instrumentalised ravaging of both inner and outer nature. The primary concern and point of alienation is science's growing instrumentalisation of conventional politics and the latter's promotion of a framework of structured silences designed to project a popular version of 'public opinion' that appears coherent and unanimously committed to techno-logistical aesthetic regimes and standards of social achievement. The mobilisation of protest action amongst the public is induced by the ensuing institutional codification of biotechnological innovation in the form of new policy and legal norms to deal with the application of bio-innovations and the risks they entail, despite uncertainties in terms of political, cultural, and scientific consensus, and the absence of a binding concept of public morality on biotechnology. Hence, the institutional codification of applications of biotechnological innovation occur largely in a state of denial of both the complex and pluralistic nature of contemporary society, and of the fundamental political rights of its people. Institutional industrial-scientific and political elites have aimed at privileging the authoritative voice of science and devaluing the non-scientist's involvement in decision-making procedures. This involves a correlative detour to a closed door policy regime with the aim of removing delicate issues relating to biotechnology from the potential reach of public challenge. A growing sub-political energy that promotes cultural, political and ethical diversity has reduced public tolerance for such institutional closure. Social movement protest politics has found its lifeblood in social and cultural unevenness and frictions, from the orientation complexes of several uncivil societies colliding. Protest politics on biotechnology issues is a reaction to the closure, stasis and repression of the scientific-industrial and political regimes. Those whom the government label 'ecophobics' and 'people-hating fanatics' have the increasing support of the public in their querying of the historically contingent social construction of nature, and our relationship to our bodies. Such actors have played a strategic role in contesting modern institutional life's 'power over' (Foucault, 1985) technologies of the self, or our field of actions and forms of thought towards inner and outer nature. This has been one of the most strain-inducing challenges posed by protest actors to the integrative capacities of the institutional support system of conventional politics, economics, and science which finds itself with little choice but to adjust to new publics demands (see Dreyer, this volume).

Science's comfortable and historically intimate relationship with institutionalised politics is currently being both exposed and problemised. Despite public dissent, the institutional discourse on biotechnology still predominantly construes democracy and justice from the perspective of the rights of science to carry out research and have the fruits of its labour approved for commercial application. There is, at present, little provision for contestation of bio-technocracy and where provision is made, it is more a case of opponents being forced into a situation of direct confrontation, or having to define themselves in terms of their distance from certain categories of normality, or what is seen to be pre-emptive. In this instance, conflict between social actors manifests itself as social discord and protest rather than discourse governed by a principle of reciprocity, with inequality of opportunity being the primary defect. Industrial and political inaction has spurred many environmental and consumer groups into direct political action. A number of protests against genetically modified food and plants in the UK have been carried out by Greenpeace, Genetic Forum (a research and lobby group), the Safe Alliance and the Consumers' Association, amongst others. A model of collective protest action relating to certain biotechnology issues has spread across national borders from Germany and the US where a series of high-profile protests and direct actions involved campaigners damaging crops and research facilities, as well as urging the public to boycott genetically modified soya and maize, and is acting as a stimulus to other groups in Britain to protest.[12] The presence of a cultural proximity means that protest actors across national contexts are united in their opposition to genetically modified food and advocate that survival does not entail a bio-revolution, as a techno-scientific rationality maintains, but an ethical one actuated in the practices of everyday life. This ethical revolution is centrally concerned with undermining the subjectification of the human/nonhuman body and acquiring a new sense of respect and responsibility towards the 'communicative body' (Frank, 1993), with its reflexive desire for dyadic socio-political and socio-cultural representation, or a critically embodied internalisation and externalisation of representations of the body. These are amongst the early signals that the gene 'revolution' will not go unchallenged and opposition seems to feed feelings of unease amongst the public at large about leaving decisions on imperatives to strategically acting bio-industrial and sympathetic policy actors. Governments have little choice but to open up the circle of actors allowed to participate in integrative negotiations directed at joint problem-solving and do so on the basis of new criteria of equality determined by social standards of relevance, rather than on the basis of considerations internal to scientific specialists. The presence of new actor types in decision-making procedures can only enhance both the institutional recognition of the (ab)normal nature of a rule-endorsed annihilation of large segments of nature and the adaptive capacities of such institutions of reflexive modernity to conditions of diversity, creativity and contradiction.

The present absence of a long-term strategic perspective on the 'why' of participation and the political role of values has been to the detriment of discursive democracy thus far, while attempts at developing new discursive political designs seem unable to deal constructively with opposition to bio-technocratic consciousness. A communicative governance, the only appropriate system of governance for the 'communicative body' (Frank, 1993) in this age of extensive societal pluralisation, will inevitably lead to decision-making procedures being fraught with a high and costly level of dissension. Undoubtedly, public participation frequently brings with it the disadvantage of decelerating decision-making and in critical moments a plea is often made for speedy, less heated solutions (see, Byk, Bora, this volume). However, the latter cannot be granted without careful consideration of the fact that the biotechnological issues being addressed today will continue to be characterised by a high level of dissension into the future. In this age of complexity, the only domain in which an ethics of responsibility on biotechnological risks can be applied is the sphere of a public that engages in critical thinking (see Strydom, this volume). Democratic will formation demands a 'dialogic politics' (Giddens, 1994; Delanty, this volume) to facilitate an open and honest discussion about what kinds of 'emancipation' in the interests of the common good (understood in the broadest pan-species sense) can genetic 'manipulation' really deliver? It is only through a more dialogically based democracy that widespread consensus on what the parameters of an ethics of collective responsibility on biotechnology could possibly emerge. The dispute over the correct paradigmatic understanding of the relationship between the legal or political system, and risk-intensive scientific innovation is primarily a political dispute, one that, in principle, affects all persons and should, therefore,

not be conducted as an esoteric discourse amongst experts. Expert systems
have an obligation to members of society who make their existence possible to respect and take on board the viewpoints of all people and not
dismiss them on the grounds they do not follow conventional criteria as to
what is the proper stuff of, say legal debate, particularly when what is currently defined as conventional is a historical construction, one that was as
controversial in previous eras as what is being newly proposed in the legal
discourse of today.

With the integration of the discursive precondition into the decision-
making process, a procedure for recognising ethical, scientific, and cultural dissent could be strengthened in the future and could potentially
reduce the likelihood of expert knowledge being inappropriately politicised and used as a strategy of evasion. Biotechnology's 'great discoveries', the accumulative achievements of linear time-bound progress, and
the genesis of a new normative and aesthetic relationship to the biological
and social body are correlative with new, more subtle political techniques
of power and control in the 'informational society' (Castells and Hall,
1994). New means of regulation – by microbiological segmentation,
seriation, synthesis, combined with macro, bureaucratically organised
classification and surveillance – have to be critically opened up as issues
for public discussion and approval. This means discussing new developments not just in terms of their scientific or immediate ethical implications
but their likely future political and commercial application in the context
of public participatory forums, the latter of which should genuinely constitute discursive intermediaries between differing claims to validity and fac-
ticity. More public discourse, however, will seem futile if there is not
complementary structural change to prevent one system's logic from un-
justly dominating. Otherwise, it will be, as always, the worldviews held by
powerful actors that continue to certify the feasibility of a new vista on the
biotechnological determination of life. For a broadly inclusive communicative governance on biotechnology issues to work effectively, two
crucial developments must ensue. First, the public must be armed with
more information on genetically-engineered products and processes than
just the current supply of alluring visions of the social nirvana said to
follow their diffusion, and secondly, a far greater degree of openness must
be shown by experts in relation to a wider interpretation of the relevant
issues. In this way, subsystems can in a more positive and critically constructive fashion become more interwoven with public discourses and vice
versa (Munch, 1992) to achieve a more complex socio-political and ethical
reasoning on biotechnology.

Notes

INTRODUCTION

1. In just one day and on one page of the *Financial Times* (13 May 1997) there is news that British Biotech, the UK's largest biotechnology company has make a significant step closer to commercialising its first product, a drug for acute pancreatis, that another UK company, Ethical Holdings is to float on the London market in response to the rapid growth in biotechnology investment in the UK and that the International Biotechnology Trust is increasing its investment in 'mid-stage' biotechnology companies to carry them from venture capital backing to a stock market listing. It is becoming apparent that investment in Europe is on the increase, though still well behind the United States, and that this investment presumes a European market from the outset which implies that regulation on an European scale will have to match this presumption.
2. The paper by John Murphy in this volume dealing with the regulatory dilemmas and strategies associated with reproductive technologies shows the complexities of regulation in a relatively accepted and advanced area of innovation.
3. See the discussions and arguments on European attempts to introduce patent protection contained in the papers by Byk and by McNally and Wheale in this volume.
4. See Dreyer this volume.
5. The proliferation of ethics commitees at both national and European level which are a response to the problems of existing legal and political institutions may be seen as a formalised though also restricted exploratory public sphere rather than a bargaining or decision-making one (see Byk below). See also Murphy (below) who shows how authorities can serve also as decision-making bodies in certain cases, hence bypassing some legitimation deficits of public decision-making.

CHAPTER 1

1. Due to limitations of space, both references and notes are kept to a minimum.
2. In his recent work, however, Habermas (1992, p. 56) takes a more sociological position insofar as he maintains that under modern conditions complex societies require that discourses be conducted permanently.
3. This cognitive dimension has steadily been gaining attention in sociology in the course of the twentieth century – from Mannheim through Habermas and Goffman to Gamson and Eder. The position adopted here owes much to Klaus Eder.

4. Apel (1991, pp. 274–5), for example, writes: 'Coresponsibility, it seems to me, is a principle of ethics that is different from, or goes beyond, the sense of justice.'
5. For instance, van Peursen (1970); Jonas (1973, 1976, 1982, 1984); Apel (1987, 1988, 1991); Melucci (1985); Beck (1988); Offe (1992); Dower (1989); Lenk (1992); Kaufmann (1992); Bernstein (1994).
6. On the question of nature at the core of the responsibility discourse, see Moscovici (1968, 1990); also Luhmann (1986), Halfmann (1986), Eder (1988, pp. 253–5), and van den Daele (1992).
7. An interesting example of the conceptual analysis of a frame is to be found in Capek's (1993) work on the 'environmental justice frame', but it is not yet sufficiently thought through. It certainly makes sense to see 'the residents' ability to mobilize for social change [as being] linked to their adoption of ... [a] ... frame', but it is misleading to identify a frame – i.e. a structure – directly with claims-making activity. In the following, I take cues from what I take to be Habermas's (1992, pp. 109–237) analysis of the rights frame but draw in particular on the work of Apel (1987, 1988, 1991) and Heller (1982, pp. 28–35).

CHAPTER 3

1. Habermas seems to have become progressively more optimistic on these issues. In *Toward a Rational Society* (1971) he stresses the dangers of scientism and technocracy. But in *The Theory of Communicative Action* (1984, 1987), and since, he has stressed the positive value of expertise in the context of highly differentiated modern societies (see, for example, 1984, pp. 71ff.)
2. The example also reveals a number of persistent patterns in the rich and strange political responses to expert advice on such risks: thus, the current (1996) treatment of British citizens of the EU is redolent of how plague victims, in earlier times, were walled into their houses.
3. This metaphor of the body as private property has come into favour in recent times. Women in particular, some pressure groups claim, own their bodies – which precludes from the start, it would seem, the thought that they are their bodies.
4. Indeed between the first and final draft of this paper IVF, and the selective abortion which often attends it, have entered the democratic discourses of everyday life with a vengeance by courtesy of the British tabloid press, and the question of what to do with your octuplets has engendered much reflexive interaction between lay and expert cultures.
5. The desire for the 'natural', and the tangible in particular, to offer us reassurances about the non-natural (the intangible, the supernatural, the moral) is by no means specific to our own 'modern', highly differentiated societies. For a brilliant discussion (and illustration) of this theme, see Spolsky (1994).
6. God and nature have long figured large in the ubiquitous social activity of passing on responsibility, but the activity has grown so popular that tech-

nical experts, bureaucrats and committees are increasingly expected to help out as well.

7. It is worth reflecting, in connection with the Warnock Report, that if those who sought to sacralise human eggs and embryos were irrational, they were but a fortnight more irrational, as it were, than the Warnock Committee itself.

8. The effect of Habermas's argument in *The Structure of Communicative Action* (1984, 1987) is to justify and valorise practically everything so far brought about in the development of modern capitalist societies, and to frame what on the face of it seem serious and far-reaching criticisms of them so that they are little more than warnings about where capitalism is in danger of going. In particular, the instrumentality he takes to be such a threat to the lifeworld is seen as innocuous, indeed as something of 'intrinsic evolutionary value' in the context of the economy (1987, p. 339). Another view is that we should recognise the differentiation and rationalisation of purely economic instrumental activity as a profoundly double-edged development and seek to overcome the division of lived experience it implies. But for Habermas such a view is mere romanticism, and as far as differentiation and rationalisation is concerned, it would seem that our present state is close to being the ideal state.

CHAPTER 4

1. The Supreme Court ruling in the case of X, a 14-year-old rape victim who sought an abortion outside the jurisdiction (1992), led to the reopening of the abortion debate in Ireland.

CHAPTER 5

1. For arguments based upon the notion of embeddedness, see especially the contributions of Granovetter (1985); Fligstein (1990); Galaskiewicz (1991); and the collection of essays edited by Powell and Di Maggio (1991). The latter also provides a sophisticated presentation and discusssion of the neo-institutionalist approach in organisational analysis. For a review of this approach in political science and sociology, see Koelble (1995).

2. These two strategies are complemented by substantial lobbying for a genetic engineering law. Business is faced with a situation in which it is in its interest to request legislation which previous lobbying aimed at preventing. Lobbying for legal institution-building is aimed at bringing to an end a situation in which an industry's performance is seriously impeded.

3. The confidence-building strategy is complemented by substantial lobbying for a deregulation of genetic engineering. This is occurring two and a half years after the July 1990 Genetic Engineering Law's enactment.

4. In the field of risk research it is above all Bayer AG which has distinguished itself. In September 1989 it initiated an expansive research project to investigate the safety of genetic engineering work.

5. The other two guidelines are environmental protection and health/safety.
6. For a developmental discussion of the communication programme, see Dreyer (1997).
7. Anton Mariacher is a member of the VCI and secretary of the *Initiative Geschützter Leben* (IGl). The IGl is a society made up of member firms and associations of the VCI. Its task is image-promotion.
8. These ads form part of a series which was launched on prominent issues concerning the chemical industry in the context of the 'Chemie im Dialog' campaign. The ads were issued in all supra-regional daily and weekly newspapers and in selected regional dailies.
9. In addition, the Hoechst AG practises 'dialogue' by *Bürgertelefon* (citizen-telephone), *Bürgerversammlungen* (citizen gatherings), and a publication series denoted 'Hoechst im Dialog'.
10. For the concept of 'öffentliche Exponiertheit' (public exposure), see Dyllick (1989, pp.15ff). According to Dyllick, indicators for public exposedness are: state intervention, political alertness, interest of science, media attention, image analyses and direct citizen participation (ibid., pp. 6–7).
11. As Poferl and Brand (1995, p. 16) have shown, as far as environmental organisations are concerned, throughout *big industry* an increased dialogic and co-operative stance has broken the dominance of strategies of polarisation. In *economic theory* as well the concept of economic rationality has increasingly been related to a concept of dialogic rationality. For a pragmatic/instrumental approach to a concept of dialogic rationality in economics see, among many others, Dyllick (1989) and Wiedmann (1992). According to Dyllick, companies today need to have a 'capability for societal self-assertion' (Dyllick, 1989, p. 477) which primarily would include three conditions and capabilities: capability for societal communication or dialogue; capability for action or co-operation; social credibility (ibid., pp. 477ff.). For a primarily normative approach which works with discourse-theoretical suppositions, see Thielemann (1994) and Ulrich (1993).
12. The theoretical perspective of the public sphere as a 'legitimatorischer Resonanzraum' of institutions is adopted from Rehberg (1995, p. 184).
13. Following Thielemann (1994, p. 123), sham co-operation is understood as a situation in which one or more but not all of the participants of an event declared as 'co-operation' or 'dialogue' merely pretend to be willing to take into consideration the claims of others according to their legitimacy.

CHAPTER 6

1. See Strydom on constructivism, this volume.
2. Rudolph (1989) uses 'connexity' as a concept to explain textual coherence or clarity and the cohesion of elements of sentences of text. Skillington (1996) extends the concept of connexity to look at textual coherence and cohesion in relation to a social and cultural context.
3. See *Irish Times*, 27 March 1995, 'The new technology', Sylvia Thompson; *Cork Examiner*, 4 January 1996, 'Gardeners angered by anti-chemical

warfare', C. T. Wilkins; *Cork Examiner*, 'Factory farming in a test tube', 9 January 1996, Denise Hall; *Irish Independent*, Paul Johnson, 'Why we must resist the temptation to play God', 8 March 1996; *Financial Times*, 'Of pigs and bioethics', Joe Rogaly, 9 March 1996.

4. *Cork Examiner*, 'Querying genetic engineering', Maureen Fox, 15 January 1993; *Irish Independent,* 'Why we must resist the temptation to play God', Paul Johnson; *Wall Street Journal*, 'Biotech under seige', John Robson, December 1996.

5. See *Earthwatch*, 'The birdless wing: Genetic engineering and the implications for farm animal welfare', Mary-Ann Bartlett: *Cork Examiner*, 'Sceptics fear scientists may unleash a cure thats worse than the disease', 21 January 1996.

6. See *Cork Examiner*, 'Will we eat up our genes from the farm in the future?' 12 January 1996, Roz Crowley.

7. See in particular *Cork Examiner*, 'Querying genetic engineering', 15 January 1993, Maureen Fox; *Cork Examiner*, 'Factory farming in a test tube', Denise Hall, 9 January 1996; *Irish Independent*, 'Why cloning animals and humans is morally wrong', Ruarc Gahan, Letters to the editor.

8. See, for example, *Irish Times*, 'The new food technology', 27 March 1995, Sylvia Thompson; *Cork Examiner*, 'Gardeners angered by anti-chemical warfare', C. T. Wilkins, 4 January 1996; *Cork Examiner*, 'Factory farming a test tube', Denise Hall, 9 January 1996; *The Sunday Times*, 'Look whats cooking', Sean Ryan, 14 May 1995.

9. See *Cork Examiner*, 'Boldly going where no ham has gone before', 20 April 1994, Valerie Robinson; *Cork Farm Examiner*, 'Biotechnology not good for all', Ann Fitzgerald, 25 August 1994; *Evening Echo*, 'The fishy tale of Loch Fyne monsters', Liam Meylin, 14 May 1995.

10. *International Environment Matters*, 'Farmers fight patents on the living', February edition, 1992.

11. See *Times Higher Education Supplement*, 'Europe's bitter harvest', Les Levidow, 11 September 1992.

12. This remark was made by Paula Giles, The Green Party's spokesperson for agriculture. Quoted in Valerie Robinson's article, 'Boldly going where no ham has gone before', *Cork Examiner*, 20 April 1994.

13. Quote taken from Steve Jones article in the *Guardian* entitled 'Don't blame the genes', 7 June 1996.

14. *International Environment Matters*, 'Farmers fight patents on the living', February edition, 1992; *Times Higher Education Supplement*, 'Europes bitter harvest', Les Levidow, 11 September 1992.

15. Quoted in Levidow, ibid.

16. See *Financial Times*, 7 March 1996; *Cork Examiner*, 'Health risk fear', 2 March 1992.

17. See *Irish Times*, 'The ear of a nude mouse bristles with ethical dilemmas', Dr William Reville, 6 November 1995; *Irish Times*, 'Scientists sound warning on sheep cloning breakthrough', 8 March 1996.

18. This quote is from Professor O'Gara from the Department of Microbiology at University College Cork, whose comments were reported in the *Cork Examiner* by Robinson, op. cit., 20 March 1994. See also *Irish Times*, 'Dealing with Genetics', Letters to the editor, 14 March 1994.

19. *Irish Times*, 'Science conference aims to ease research fears', Dick Horgan, 7 January 1995. In support of this sentiment, see also *Irish Times*, 'Alzheimer's cure might arise from research on mice', Mary Mulvihill, 23 February 1995.

20. *Cork Examiner*, 'UCC tests on live animals', Ailin Quinlan, 11 February 1993; *Wall Street Journal*, 'Biotech under seige', John Robson, 1996.

21. See *Cork Examiner*, Robinson, op. cit., 20 April 1994; *Irish Times*, 'Gene proposal killed off', Patrick Smith, 2 March 1995.

22. Professor O'Gara from the Department of Microbiology in University College, Cork, quoted by Peter Gleeson in *Cork Examiner* ('Friendly bacteria to fight disease in crops', 28 September 1994); *Irish Farmers Monthly*, 'It's all in the genes', Matt O'Keeffe, November 1995.

23. See *Cork Examiner*, Gleeson, op. cit., 28 September 1994.

24. *Irish Times*, 'Killer cauliflowers prepare to bite back', Mary Mulvihill, 24 January 1995.

25. Quoted in *The Sunday Times*, 'Look what's cooking', Sean Ryan, 14 May 1995.

26. *Irish Times*, 'UCC to coordinate major EC project', Des O'Sullivan, 3 September 1995; *Irish Press*, 'Ireland has edge in biotechnology', Pat Holmes, 1 February 1993; *Financial Times*, 'High tech sheep coming to market for £100 million listing', David Green, 3 May 1996.

27. See *Wall Street Journal*, 'Biotech under siege', John Robson, December 1995.

28. See *Sunday Business Post*, 'A land of milk and money', Helen Callanan, 25 April 1993; *Irish Times*, 'Tomato war foreseen as biotechnology delivers its first whole food', Conor O'Clery, 20 May 1994; *Irish Times*, Mulvihill, op. cit., 24 January 1995; *Financial Times*, 'Mice offer key to Alzheimer's cure', Clive Cookson, 7 March 1996; *Daily Telegraph*, 'Sheep tonic takes medicine into a new field' Auslan Cramb, 4 May 1996.

29. *Irish Farmers Monthly*, O'Keeffe, op. cit., November 1995.

30. Qouted in the *Cork Examiner*, 'UCC test on live animals', Ailin Quinlan, 11 February 1993.

31. See *Farm Examiner*, 'Biotech diet of the future', Stephen Codagon, 18 January 1996.

32. See *Irish Times*, op. cit., O'Clery, 20 May 1994; *Cork Examiner*, Peter Gleeson, op. cit., 28 September 1994.

33. Kitschelt (1986) defines opportunity structures as comprising institutional arrangements, the historical antecedents for political mobilisation, as well as the social and cultural climate.

CHAPTER 7

1. Cloning, for example, is almost universally regarded as an unacceptable enterprise. English law prohibits it *absolutely* as regards human beings: Human Fertilisation and Embryology Act 1990, s. 3(3)(d). (Hereafter, 1990 Act.)

2. Hereafter, IVF.

3. Other, related tensions are addressed elsewhere in this volume.

4. Notably, the Royal Commission on New Reproductive Technology in Ottowa has proposed legislation designed to criminalise many commercial forms of assisted conception remunerated sperm donation.

5. See further G. Douglas and N. Lowe, 'Becoming a Parent in English Law' (1992) 108 *Law Quarterly Review* 414.

6. That is, male sperm or female ova.

7. Ibid., ch. 6. But note that these statistics are apt to mislead on the breadth of demand since, in the UK, section 13(5) of the 1990 Act makes it probable that IVF is provided only to heterosexual *couples*. Thus the statistics do not include single and lesbian women who desire assisted conception.

8. In 1992 there were approximately 12,000 IVF cycles and 30,000 donor insemination cycles: see D. Morgan, 'What HFEA did in its first year' [1992] *Bulletin of Medical Ethics* 17, at p. 18.

9. Infertility is a slippery concept for which there is no universally accepted meaning: (see Douglas 1991, pp. 103–6).

10. For example, lesbianism. In such a case, however, the debate acquires a further dimension: whether there is any moral basis for 'playing God' where, biologically, there is no need to do so.

11. For example, blocked fallopian tubes.

12. 'Reproductive technologies' is a convenient phrase; it excludes non-biotechnological forms of assisted conception such as 'basic surrogacy' (i.e. sexual intercourse between the commissioning father and the surrogate mother) and 'low-tech surrogacy' (i.e. artificial insemination of the surrogate mother with the commissioning father's sperm).

13. Emphasis supplied. We must, however, distinguish two senses of the right to reproduce. According to the first of these, the right is characterised by a claim (against the state) to be provided with treatment services wherever needed or desired. According to the second, the right holder possesses no more than a *liberty* to reproduce according to whatever method s/he chooses. Under English law, there is no right to infertility treatment services in the first sense: *R v Ethical Committee of St Mary's Hospital ex p. Harriott* [1988] 1 FLR 512.

14. This debate did *not* figure at all in the Report produced by the Committee of Inquiry which was set up to consider the 'social, ethical and legal implications' associated with 'recent and potential developments in medicine and science related to human fertilisation and embryology'. Rather, the view expressed was that 'if an embryo is human and alive, it follows that it should not be deprived of a chance for development' (DHSS, 1984 para. 11.9).

15. Cf. the view that, at most, there should be 'presumptions in law, of rights to apply for treatment and services by aspiring parents' with the facility for 'refusals [of those treatments] based on published rules, subject to appeal or review in exceptional cases'. (See Craven and Sood, 1988, p. 467.)

16. This has been recognised in England in *Re D* [1976] 1 All ER 326 *per Heilbron J* at 332 and in Canada in *Re Eve* (1986) 31 DLR (4th) 1 *per* La Forest J at 5.

17. Some, however, might argue that there *is* a public interest at stake: the potential affront to the public conscience. But how can the creation

of embryonic human life be more repugnant than its destruction and how, therefore, could one seek to proscribe IVF whilst abortion remains lawful?

18. This syllogism presupposes that the commodification of human tissue is not morally repugnant and hence a basis for legal proscription. It has been argued, for instance, that 'if an entity [e.g. a human ovum or sperm] has dignity, then treating that entity or some part of it as a commodity is morally objectionable if the treatment offends dignity' (Munzer, 1993). For a contrary view see Duxbury (1996). See also DHSS (1987).

19. 1990 Act, s. 3(3)(d).

20. Ibid., s. 3(3)(b).

21. Ibid., Shed. 2, para 3(5).

22. This regulatory technique has long existed in English law: section 1 of the Abortion Act 1967 stipulates that only qualified medical practitioners may lawfully terminate a woman's pregnancy.

23. But note that under the 1990 Act, only properly licensed clinics are permitted to make available 'treatment services'.

24. There could be as many as three women with a claim to motherhood: the genetic parent, gestational parent and the social parent. This could happen where A donates eggs used to produce a pregnancy in B who was to act as a surrogate mother for C.

25. In both *A* v *C* (1978) 8 *Fam Law* 170 and *P* v *B* (1987) *The Times*, 13 March the surrogate mother changed her mind about handing over the children.

26. In England, for example, the Child Support Act 1991 obliges all absent parents to make financial provision, in respect of their children.

27. Child Support Act 1991, s. 1. See further Jacobs and Douglas (1993).

28. For example, the right to choose the child's name and the right to consent to medical treatment. See further Hoggett (1993, pp. 10–23).

29. 1990 Act, ss 27–30.

30. In *Re W* [1992] 3 WLR 758 it was held that parents had the power to consent to a whooping cough inoculation despite it not being a therapeutic procedure.

31. In Sweden, where the right to remain anonymous was removed, the number of donors declined dramatically.

32. The Hippocratic Oath states: 'All that may come to my knowledge in the exercise of my profession ... which ought not to be spread abroad, I will keep secret and never reveal'. See McHale (1992).

33. In the present context, fatherlessness extends beyond estrangement or death of a father. Under the 1990 Act, s. 28, a child may be treated as legally fatherless where she was conceived by use of donated sperm. This 'is intended to afford protection to a donor whose sperm is used in accordance with his consent to establish a pregnancy'. See Morgan and Lee (1991, p. 156).

34. For a powerful counter-argument, see K. O'Donovan, 'A right to know one's parentage?' (1988) 2 *International Journal of Law and the Family* 27.

35. This would occur if a couple, A and B, used the sperm (or ovum) of their friend C in order to have a child.

36. *Gaskin* v *United Kingdom* [1990] 1 FLR 167.

37. 1990 Act, s. 31(3), (4). To facilitate this, the Human Fertilisation and Embryology Authority must keep a record of, *inter alia*, 'any identifiable individual [who] was, or may have been, born in consequences of treatment services': Ibid., s. 31(1), (2).

38. Ibid., s 31(4)(b). The concern here is grounded in eugenics: to avoid the prospect of a marriage that would fall within the prohibited degrees of consanguinity (because of the risk that any children would be born with disabilities).

39. *Gaskin* v *United Kingdom*, *supra*, n 37.

40. 1990 Act, s. 31(5).

41. See J. Triseliotis (1984) on the Adoption Act 1976, s. 51 which confers the related right to obtain a copy of one's original birth certificate.

42. For an excellent introductory account, see Grubb (1988).

43. 'Treatment services' is the shorthand expression adopted in s. 3 of the 1990 Act to describe IVF, GIFT (gamete intrafallopian transfer), ZIFT (zygote intrafallopian transfer), etc.

44. In *Emeh* v *Kensington and Chelsea and Westminster AHA* [1984] 3 All ER 1044 it was held that a wrongful birth action could brought by a woman who became pregnant because of the defendant's failure properly to perform a sterilisation operation.

45. [1982] 2 All ER 777.

46. Ibid., at 781.

47. See further Murphy (1994).

48. The Supply of Goods and Services Act 1982, s. 13 implies into contracts for services the requirement that those services will be performed with reasonable care and skill.

49. In principle, the plaintiff could call expert evidence but there are two frequent problems even here. First, expert evidence is costly to mobilise; secondly for every expert who will argue X for the plaintiff, there is usually another expert who can argue Y for the defence.

50. For a contrary view, see Duxbury (1996, n 20).

51. Such licences are available under s. 11(1) of the 1990 Act.

52. 1990 Act, s. 9(8).

53. The solution is partial in that even if IVF allows a couple to have a child, it does not *cure* the underlying infertility. See Douglas (1991, pp. 106–10).

54. See Brazier (1989) and Harris (1990).

55. For the various arguments of principle, see Morgan and Lee (1991).

56. See Montgomery (1991).

57. Ibid. There is a limited exception to this rule in Sched. 2, para. 5 which provides for licences to 'authorise the mixing of [human] sperm with the egg of a hamster' for the sole purpose of 'developing more effective techniques for determining the fertility or normality of sperm' and then 'only where anything which forms is destroyed when the research is complete and, in any event, not later than the two-cell stage'.

58. Ibid., Sched 2, para. 3(2).

59. 1990 Act, s. 7.

60. Under ss 23 and 24 of the 1990 Act, apart from the power to issue licences, the Authority is able to issue non-derogable directions to practitioners, failure to comply with which entitles the Authority to revoke the licence.

61. 1990 Act, Sched 1, para. 4.
62. Ibid., Sched 1, para. 1.
63. In its present composition, the Authority has only one theologian and one specialist ethicist.
64. 1990 Act, Sched 1, para. 4(1).
65. But note the caveat that medico-moral issues tend only partially to divide politicians along traditional party lines.
66. Bodies of this kind are not new, witness, e.g. the Health and Safety at Work Executive which polices specialist employment legislation.

CHAPTER 8

1. An especially telling example was provided when environmental groups withdrew from the WZB-sponsored technology impact assessment of herbicide-resistant plants just two days before it ended (see van den Daele, 1994). Spangenberg (1993) spoke generally of 'participation overkill'.
2. On the situation in France, see Lafaye and Thevénot (1993) and Defrance (1988). The Ministry of Housing (1994) gave an overview of the legal situation in the EU member states.
3. The case of participatory technology impact assessment has been treated in Bora (1995), Bora and Döbert (1993), Döbert (1994a) and Daele (1994).
4. Of course, a certain degree of restriction stems from the fact that 'only' procedural control is granted, not discretionary control (see Bora, 1994).
5. See Döbert (1994a) on the social significance of different environmental ethics.
6. This refers directly to Derrida's critique of Bejamin, cf. Benjamin (1921/1965), Derrida (1990, 1994).
7. On these points, see the criticisms in Eder (1989).

CHAPTER 9

1. Introductory comments to Workshop on 'Biotechnological Innovation, Societal Responses, Policy Implications', held in Cork in March 1995 by the editor of this volume.

CHAPTER 11

1. See 'Cloning presents an opportunity not a threat', Editorial Commentary. *The Independent*, 28 February 1997.
2. In Britain, government has cut more than £1 billion from research funding over the last decade (see 'When the price is wrong', Stephanie Pain, *The Guardian*. 27 February 1997).
3. See 'Suspicion of science based on wishful thinking', Dr William Reville. *The Irish Times*, 2 September 1996.

4. See Charles Arthur ('First cloned lamb paves way for life by production line', *The Independent*, 24 February 1997), who quotes Dr Ian Wilmut of the Roslin Institute for his views on human cloning.

5. Stephen Bloom, Professor of Endocrinology at the Hammersmith Hospital in London believes that approximately 50 per cent of obesity possesses a genetic root. This leads to the belief that there may be genetic treatments for obesity, known as the ob, agouti, tub, db and fat genes (see 'Genetics wrestles with obesity', Daniel Green, *Financial Times*, 30 May 1996).

6. For a discussion of collective learning processes, see Piet Strydom (1987), 'Technology as a Collective Learning Process'. Paper presented to the Annual Conference of the Sociological Association of Ireland, Dublin, 24–25 April; and Patrick O'Mahony (1991), 'Science and Industry', *Education and Training Technology International*, Vol. 28, No 1.

7. See 'Continuing scandal of the food we eat', Catherine Bennett. *The Guardian*, 26 June 1996.

8. See 'Cold comfort farming', Graham Harvey, *The Observer*, 9 March 1997.

9. The European Union has now officially given the go-ahead for genetically modified maize produced in the US by chemical giant Ciba-Geigy to be imported into Europe. This maize contains a bacterial marker gene which increases the plants resistance to pests and disease, but also more controversially, to the antibiotic ampicillin used in animal and human medicines. This latter characteristic has caused concern for environmental groups who point out that the bacterial marker gene could enter the food chain through animal feeds. The European Union has postponed the issue of whether modified maize should be labelled until next year.

10. For a more detailed analysis, see 'Mad cows and Irishmen', Emmet Oliver, *The Irish Times*, 22 January 1997.

11. See 'Time for the great leap backward', Hugo Young. *The Guardian*, 4 June 1996.

12. See 'Crop scientists hit by green action', Peter Beaumont, *The Observer*, 15 December 1996.

Bibliography

Abbott, A. (1995a), 'EPO Verdict Lifts Patent Coverage of Seeds', *Nature,* 374, 2 March, p. 8.

Abbott, A. (1995b), 'European Proposal Reopens Debate over Patenting of Human Genes', *Nature*, 378, 21/28 December, p. 756.

Abbott, A. (1996), 'Withdrawal of Patent Claim Leaves Position of Plants Unclarified', *Nature,* 381, May, p. 178.

Adorno, T. and Horkheimer, M. (1979), *The Dialectic of Enlightenment* (London: NLB).

Aglietta, M. (1979), *A Theory of Capitalist Regulation: the United States Experience* (Translated by D. Fernbach) (London: NLB).

Altimore, M. (1980), 'The Social Construction of a Scientific Controversy: Comments on Press Coverage of the Recombinant DNA Debate', *Science, Technology and Human Values*, 7, 41.

Anderson, C. (1993), 'While CDC Drops Indian Tissue Claim', *Science,* 26, 25 November, p. 831.

Apel, K. O. (1978), *Diskurs und Verantwortung: Das Problem des Übergangs zur postkonventionellen Moral* (Frankfurt: Suhrkamp).

Apel, K. O. (1980), 'The *A Priori* of the Communication Community and the Foundation of Ethics: the Problem of a Rational Foundation of Ethics in the Scientific Age', in *The Transformation of Philosophy* (London: Routledge and Kegan Paul).

Apel, K. O. (1984), 'The Situation of Humanity as an Ethical Problem', in *Praxis International*, 4, 3, pp. 250–65.

Apel, K. O. (1987), 'The Problem of a Macroethics of Responsibility to the Future in the Crisis of Technological Society', *Man and World*, 20, pp. 3–40.

Apel, K. O. (1988), 'The Conflicts of Our Time and the Problem of Political Ethics', in F. R. Dallymar (ed.) *From Contract to Community: Political Theory at the Crossroads* (New York: Dekker).

Apel, K. O. (1988), *Diskurs und Verantwortung* (Frankfurt: Suhrkamp).

Apel, K. O. (1991), 'A Planetary Macroethics for Mankind', in E. Deutsch (ed.) *Culture and Modernity* (Honolulu: University of Hawaii Press).

Ashby, W. R. (1956), *An Introduction to Cybernetics* (London: Chapman and Hall).

Bakhtin, M. (1973), *Problems in Dostoevsky's Poetics* (Ann Arbor: Ardis).

Bakhtin, M. (1981), 'Discourse in the Novel', in *The Dialogic Imagination: Four Essays by Mikhail Bakhtin* (translated by C. Emerson and M. Holoquist) (Austin: University of Texas Press).

Barnes, B. (1995), *The Elements of Social Theory* (London: UCL Press).

Barthe, S. and Dreyer, M. (Project Managers: K. Eder and K. W. Brand) (1995), 'Reflexive Institutionen? Eine Untersuchung zur Herausbildung eines neuen Typus Institutioneller Regulungen im Umweltbereich', *Interim Report of the DFG Project* (München: Munchner Projektgruppe für Sozialforschung).

Beck, U. (1988), *Gegengifte* (Frankfurt: Suhrkamp).

Beck, U. (1992), *Risk Society: Towards a New Modernity* (London: Sage).

Beck, U. (1995a), *Ecological Politics in an Age of Risk* (Cambridge: Polity).

Beck, U. (1995b), *Ecological Enlightenment: Essays on the Politics of the Risk Society* (Englewood Cliffs, NJ: Humanities Press).

Beck, U. (1996), 'World Risk Society as Cosmopolitan Society? Ecological Questions in a Framework of Manufactured Uncertainties', *Theory, Culture and Society*, 13, 4.

Benjamin, W. (1965), *Zur Kritik der Gewalt und andere Aufsätze* (Frankfurt: Suhrkamp) (original work published 1921).

Bernstein, R. (1994), 'Rethinking Responsibility', *Social Research* 61, 4, pp. 833–52.

Bijker, W. E. et al. (eds) (1990), *The Social Construction of Technological Systems* (London: MIT).

Blaug, R. (1996), 'New Theories of Discursive Democracy', *Philosophy and Social Criticism*, 22, 1, pp. 49–80.

Blumer, H. (1966), 'The Mass, the Public, the Public Opinion', in B. Berelson and M. Jannowitz (eds) *Reader in Public Opinion and Communication* (New York: Free Press, 2nd edition).

Bohme, G. van den Daele, W. and Krohn, W. (1978), 'The Scientification of Technology', in W. Schafer (ed.) *Finalisation in Science* (Dordrecht: D. Reidel).

Bora, A. (1994), 'Schwierigkeiten mit der Öffentlichkeit zum Wegfall des Erörterungstermins bei Freisetzungen nach dem novellierten Gentechnikgesetz', *Kritische Justiz*, 3, pp. 306–22.

Bora, A. (1995), 'Procedural Justice as a Contested Concept: Sociological Remarks on the Group Value Model', *Social Justice Research*, 2, pp. 175–95.

Bora, A. and Dobert, R. (1993), 'Konkurrierende Rationalitaten: Politischer und technisch-wissenschaftlicher Diskurs im Rahmen einer Technikfolgenabschatzung von gentechnisch erzeugter Herbizidresistenz in Kulturpflanzen, *Soziale Welt*, 44, 1, pp. 75–97.

Bourdieu, P. (1992), 'Thinking about Limits', in M. Featherstone (ed.) *Cultural Theory and Cultural Change* (London: Sage).

Bourdieu, P. (1974), *Zur Soziologie der symbolischen Formen* (Frankfurt: Suhrkamp).

Braun, E. (1980), *Wayward Technology* (London: Frances Pinter).

Brazier, M. (1989), '"Embryos' Rights": Abortion and Research', in M. D. A. Freeman (ed.) *Medicine, Ethics and the Law* (London: Stevens).

Buchel, K. H. (1989), 'Gentechnik bei Bayer fur Medzin und Landwirtschaft', in A. G. Bayer (ed.) *Gentechnik bei Bayer* (Presse: forum) 27–28 September, pp. 12–29 (Leverkusen: Bayer A. G).

Butler, D. (1995), 'Genetic Diversity Proposal Fails to Impress International Ethics Panel', *Nature*, 377, 5 October, p. 373.

Butler, D. and Gershon, D. (1994), 'Breast Cancer Discovery Sparks New Debate on Patenting Human Genes', *Nature*, 371, 22 September, pp. 271–2.

Calhoun, C. (ed.) (1992), *Habermas and the Public Sphere* (Cambridge, Mass: MIT Press).

Capek, S. M. (1993), 'The "Environmental Justice" Frame', *Social Problems*, 40, pp. 5–24.

Castells, M. and Hall, P. (1994), *Technopoles of the World* (London: Routledge).

CEC (1988), 'Proposal for a Council Directive on the Legal Protection of Biotechnological Inventions' (Brussels: European Commission), COM (88) final -SYN 159, October 17th.

CEC (1991), 'Promoting the Competitive Environment for the Industrial Activities Based on Biotechnology within the Community' (Brussels: European Commission) III A 3, 15 April.

CEC (1992), 'Official Journal: European Patent Office, 10, p. 588.

CEC (1994), *Growth, Competitiveness, Employment: the Challenges and Ways Forward into the 21st Century* (Brussels: European Commission).

CEC, Council of Ministers (1994), 'Common Position (EC) No. 4/94 adopted by the Council on 7 February 1994 with a View to Adopting European Parliament and Council Directive on the Legal Protection of Biotechnological Inventions', *Official Journal,* No. C 101, pp. 65–75, 9 April.

CEC (1995), 'Proposal for a European Parliament and Council Directive on the Legal Protection of Biotechnological Inventions', *European Commission,* COM(95) 661 final, 95/0350 (COD), 13 December.

Clausen, L. and Dombrowsky, W. R. (1984), 'Warnpraxis und Warnlogik', *Zeitschrift für Soziologie,* 13, pp. 293–307.

Collingridge, D. (1980), *The Social Control of Technology* (London: Frances Pinter).

Craven G. J. and Sood, U. (1988), 'Becoming a Parent: Decisions, Rights and Legal Issues', *Family Law,* 463.

Crowe, C. (1990), 'Whose Mind over whose Matter? Women, in Vitro Fertilization and the Development of Scientific Knowledge', in M. McNeil et al. (eds) *The New Reproductive Technologies* (London: Macmillan).

Daele, W. van den (1986), 'Technische Dynamik und gesellschaftliche Moral', *Soziale Welt,* 37, pp. 149–72.

Daele, W. van den (1990), 'Risiko-Kommunikation: Gentechnologie', In H. Jungermann, B. Rohrmann, and P. M. Wiedemann (eds), *Risiko-Konzepte, Risiko-Konflikte, Risiko-Kommunikation* (Julich: Forschungszentrum Julich).

Daele, W. van den (1992), 'Concepts of Nature in Modern Societies and Nature as a Theme in Sociology', in M. Dierkes and B. Biervert (eds) *European Social Science in Transition* (Frankfurt: Campus).

Daele, W. van den (1994), 'Technology Assessment: a Political Experiment', *WZB Discussion Paper,* pp. 92–319 (Berlin: Wissenschaftszentrum Berlin für Sozialforschung).

Defrance, J. (1988), '"Donner" la Parole: La construction d'une relation d'echange', *Actes de la Recherche en Sciences Sociales,* 73, pp. 53–65.

Department of Health, Ireland (1986), *Report on Health Services 1983–1986* (Dublin: Government Stationery Office).

Department of Health and Social Security, United Kingdom (1984), *Report of the Committee of Inquiry into Human Fertilisation and Embryology,* Cmnd 9314 (London: DHSS).

Department of Health and Social Security, United Kingdom (1987), *Human Fertilisation and Embryology: a Framework for Legislation* (London: DHSS, Cm 259).

Derrida, J. (1990), 'Deconstruction and the Possibility of Justice', *Cardozo Law Review,* 11, pp. 5–6.

Derrida, J. (1994), *Spectres of Marx: The State of Debt, the Work of the Mourning, and the New International* (translated by P. Knauf) (New York: Routledge).

Dewey, J. (1927), *The Public and its Problems* (New York: Henry Holt).

Dickson, D. (1994), 'HGS Seeks Exclusive Option on all Patents Using its rDNA Sequences', *Nature,* 371, 6 October, p. 463.

Dickson, D. (1995), 'Open Access to Sequence Data "Will Boost Hunt for Breast Cancer Gene"', *Nature,* 378, 30 November, p. 425.

Dickson, D. (1996), 'Whose Genes Are They Anyway?', *Nature,* 381, 2 May, pp. 11–14.

Dickson, D. and Jayaraman, K. S. (1995), 'Aid Groups Back Challenge to Name Patents', *Nature,* 377, 14 September, p. 95.

Döbert, R. (1994a), 'Handlungs-Partizipationskosten und die Reproduktion neokonstruktivistischer Relativismen: Ein Blick auf ein erhellendes Ende einer Technik-folgenabschatzung' (Wissenschaftszentrum Berlin für Sozialforschung, Unpublished MS).

Döbert, R. (1994b), 'Die Überlebenschancen unterschiedlicher Umweltethiken', *Zeitschrift für Soziologie,* 23, pp. 306–22.

Douglas, G. (1991), *Law, Fertility and Reproduction* (London: Sweet and Maxwell).

Douglas G. and Lowe, N. (1992), 'Becoming a Parent in English Law', *Law Quarterly Review,* 108.

Douglas, M. (1990), 'Risk as a Forensic Resource', *Daedalus,* 199, 4, pp. 1–16.

Dower, N. (1989), *Ethics and Environmental Responsibility* (Aldershot: Avebury).

Dreyer, M. (1997), 'Die Kommunikationspolitik der Chemischen Industrie im Wandel', In K. W. Brand, K. Eder and A. Poferl (eds), *Okologische Kommunikation in Deutschland,* (Opladen: Westdeutscher Verlag).

Dryzek, J. (1990), *Discursive Democracy: Policy, Politics and Political Science* (Cambridge: Cambridge University Press).

Duxbury, N. (1996), 'Do Markets Degrade?', *Modern Law Review,* 59.

Dyllick, T. H. (1989), *Management der Umweltheziehungen. Offentliche Auseinandersetzungen als Herausforderung* (Weisbaden: Gabler).

Eder, K. (1988), *Die Vergesellschaftung der Natur* (Frankfurt: Suhrkamp).

Eder, K. (ed.) (1989), 'Klassenlage, Lebensstil und kulturelle Praxis: Theoretische und empirische Beitrage zur Auseinandersetzung mit Pierre Bourdieus Klassentheorie (Frankfurt: Suhrkamp).

Eder, K. (1992), *Framing and Communicating Environmental Issues: a Discourse Analysis of Environmentalism* (European University Institute, Florence, Unpublished MS).

Eder, K. (1993), *The New Politics of Class* (London: Sage).

Eder, K. (1996), *The Social Construction of Nature* (London: Sage).

Eder, K. (1996a), 'The Institutionalisation of Environmentalism: Ecological Discourse and the Second Transformation of the Public Sphere', in S. Lash et al. (eds) *Risk, Environment and Modernity* (London: Sage).

Emmott, S. (1996), *The Case against Patents in Genetic Engineering* (London: Genetics Forum).

Evers, A. and Nowotny, H. (1987), *Über den Umgang mit Unsicherheit* (Frankfurt: Suhrkamp).

Fairclough, Norman (1992), *Discourse and Social Change* (Cambridge: Polity Press).

Feher, F. and Heller, A. (1994), *Biopolitics* (Aldershot: Avebury).

Fligstein, N. (1990), *The Transformation of Corporate Control* (Cambridge, Mass: Harvard University Press).

Foucault, M. (1977), *Die Ordnung des Diskurses: Inauguralvorlesung am College de France, Dezember 2nd, 1970* (Frankfurt: Ullstein).

Foucault, M. (1985), *The History of Sexuality, Vol. 2* (New York: Pantheon).

Fowler, C. and Mooney, P. (1990), *The Threatened Gene: Food, Politics, and the Loss of Genetic Diversity* (Cambridge: Lutterworth Press).

Frank, A. W. (1993), 'For a Sociology of the Body: an Analytical Review', in M. Featherstone, M. Hepworth and B. S. Turner (eds) *The Body* (London: Sage).

Franklin, S. (1990), 'The Social Construction of Infertility', in M. McNeil et al. (eds) *The New Reproductive Technologies* (London: Macmillan).

Fraser, N. (1989), *Unruly Practices: Power, Discourse and Gender in Contemporary Social Theory* (Cambridge: Polity Press).

Galaskiewicz, J. (1991), 'Making Corporate Actors Accountable: Institution-building in Minneapolis-St Paul', In W. W. Powell and P. J. DiMaggio (eds) *The New Institutionalism in Organisational Analysis* (Chicago: University of Chicago Press).

Giddens, A. (1990), *The Consequences of Modernity* (Cambridge: Polity Press).

Giddens, A. (1994), *Beyond Left and Right* (Cambridge: Polity Press).

Giegel, H. (1992), 'Diskursive Verstandigung und systemische Selbssteurung', in H. Giegel (ed.) *Kommunikation und Konsens in modernen Gesellschaften* (Frankfurt: Suhrkamp).

Gouldner, A. D. (1976), *The Dialectic of Ideology and Technology* (London: Macmillan).

GRAIN (1990), *Disclosures* (Barcelona: ICDA & GRAIN).

Granovettor, M. (1985), 'Economic and Social Structure: The Problem of Embeddedness', *American Journal of Sociology,* 91, 3, 481–510.

Grant, W., Martinelli, A. and Paterson, W. (1989), 'Large Firms as Political Actors: a Comparative Analysis of the Chemical Industry in Britain, Italy and West Germany', *Western European Politics,* 12, 2, pp. 72–90.

Grubb, A. (1988), 'Conceiving: a New Cause of Action', in M. D. A. Freeman (ed.) *Ethics and the Law* (London: Stevens).

Haas, Ernst, B. (1990), *When Knowledge is Power: Three Models of Change in International Organisations* (Berkeley: University of California Press).

Habermas, J. (1971), *Toward a Rational Society* (London: Heinemann).

Habermas, J. (1974), *Theory and Practice* (London: Heinemann).

Habermas, J. (1984), *The Theory of Communicative Action, Vol. 1* (London: Heinemann).

Habermas, J. (1987), *The Philosophical Discourse of Modernity* (Cambridge: Polity).

Habermas, J. (1987), *The Theory of Communicative Action, Vol. 2* (Oxford: Polity).

Habermas, J. (1989), *The Structural Transformation of the Public Sphere* (Cambridge, Mass: MIT).

Habermas, J. (1990), *Justification and Application* (Cambridge: Polity).

Habermas, J. (1992), *Faktizität und Geltung* (Frankfurt: Suhrkamp).

Habermas, J. (1993), *Justification and Application: Remarks on Discourse Ethics* (Cambridge, Mass: MIT).

Habermas, J. (1994), 'Three Normative Models of Democracy', *Constellations*, 1, 1, pp. 1–10.

Habermas, J. (1995), 'On the Internal Relationship Between the Rule of Law and Democracy', *European Journal of Philosophy*, 3, 1, pp. 12–20.

Habermas, J. (1996), *Between Facts and Norms: Contributions to a Discourse Theory of Law and Democracy* (translated by William Rehg) (Cambridge: Polity).

Hajer, M. A. (1995), *The Politics of Environmental Discourse: Ecological Modernisation and the Policy Process* (Oxford: Clarendon).

Halfmann, J. (1986), 'Autopoiesis und Naturbeherrschung', in H. J. Unverferth (ed.) *System und Selbstproduktion* (Frankfurt: Lang).

Hall, S., Critcher, C., Jefferson, T., Clarke, J. and Roberts, B. (1978), *Policing the Crisis: Muggings, the State and Law and Order* (London: Macmillan).

Halliday, M. A. K. (1978), *Language as Social Semiotic: the Social Interpretation of Language and Meaning* (London: Edward Arnold).

Halliday, M. A. K. (1985), *An Introduction to Functional Grammar* (London: Edward Arnold).

Hannigan, J. (1995), *Environmental Sociology: a Social Constructionist Perspective* (London: Routledge).

Haraway, D. (1985), 'A Manifesto for Cyborgs: Science, Technology and Socialist Feminism in the 1980's', *Socialist Review,* 80, pp. 64–107.

Harraway, D. (1995), 'Otherworldly Conversations, Terran Topics, Local Terms', in V. Shiva and I. Moser (eds) *Biopolitics* (London: Zed Books).

Harris, J. (1990), 'Embryos and Hedgehogs: On the Moral Status of the Embryo', in A. Dyson and J. Harris (eds) *Experiments on Embryos* (London: Routledge).

Harrison, R. F. et al. (1992), 'An Irish Out-patient-Based In-vitro Fertilisation Service', *Irish Medical Journal,* 85, 2, pp. 63–65.

Heller, A. (1982), *A Theory of History* (London: Routledge and Kegan Paul).

Hilgartner, S. (1995), 'The Human Genome Project', in S. Jasanoff (ed.) *Handbook of Science and Technology Studies* (London: Sage).

Hoggett, B. (1993), *Parents and Children* (London: Sweet and Maxwell).

Initiative 'Geschutzler leben' (ed.) (1991), 'Infobrief, 2/91' (Frankfurt).

Initiative 'Geschutzler leben' (ed.) (1992), *Chemie im Dialog: Mitgliedersammlung 1992* (Frankfurt).

Initiative 'Geschutzler leben' (ed.) (1993), *Chemie im Dialog: Mitgliedersammlung 1992* (Frankfurt).

Initiative 'Geschutzler leben' (ed.) (1994), *Chemie im Dialog: Mitgliedersammlung 1992* (Frankfurt).

Jacobs E. and Douglas, G. (1993), *Child Support: the Legislation* (London: Sweet and Maxwell).

Jayaraman, K. S. (1996), 'Indian Researchers Press for Stricter Rules to Regulate "Gene Hunting"', *Nature,* 379, 1 February, pp. 381–2.

Jonas, H. (1973), 'Technology and Responsibility: Reflections on the New Tasks of Ethics', *Social Research* 40, 1, pp. 31–54.

Jonas, H. (1974), 'Technology and Responsibility: Reflections on the New Tasks of Ethics', in *Philosophical Essays: From Ancient Creed to Technological Man* (Englewood Cliffs, NJ: Prentice-Hall).

Jonas, H. (1976), 'Responsibility Today: the Ethics of an Endangered Future', *Social Research* 43, 1, pp. 77–113.

Jonas, H. (1982), 'Technology as a Subject for Ethics', *Social Research*, 49, pp. 891–98.

Jonas, H. (1984), *The Imperative of Responsibility: In Search of an Ethics for the Technological Age* (Chicago: University of Chicago Press).

Jonas, H. (1984), *Das Prinzip Verantwörtung* (Frankfurt: Suhrkamp).

Jonas, H. (1994), 'Philosophy at the End of the Century: a Survey of its Past and Future', *Social Research*, 61, 4.

Jordanova, L. J. (1980), *Natural Facts: a Historical Perspective on Science and Gender* (Cambridge: Cambridge University Press).

Juma, C. (1989), *The Gene Hunters* (London: Zed Books).

Kaufmann, F. X. (1992), *Der Ruf nach Verantwörtung* (Freiburg: Herder).

Kesselring, S. (1995), 'Die Storfalle bei Hoechst im Fruhjahr 1993 – Eine diskusanalytische Fallstudie' München, Diploma Thesis (Ludwig Maximilian-Universität München).

Kingdom, E. (1991), *What's Wrong with Rights? Problems for Feminist Politics of Law* (Edinburgh: Edinburgh University Press).

Kitschelt, H. (1986), 'Political Opportunity Structures and Political Protest: Anti-Nuclear Movements in Four Democracies', *British Journal of Political Science*, 16, pp. 57–85.

Koelble, T. A. (1995), 'The New Institutionalism in Political Science and Sociology', *Comparative Politics*, 27, 1, pp. 231–43.

Kollek, R. (1995), 'The Limits of Experimental Knowledge: a Feminist Perspective on the Ecological Risks of Genetic Engineering' in V. Shiva and I. Moser (eds) *Biopolitics* (London: Zed Books).

Kreibich, R. (1986), *Die Wissenschaftsgesellschaft* (Frankfurt: Suhrkamp).

Krimsky, S. (1992), 'Regulating Recombitant DNA Research and Its Applications', in D. Nelkin (ed) *Controversy* (London: Sage).

Lafaye, C. and Thevenot, L. (1993), 'Une Justification Ecologique?: Conflicts dans l'amènagement de la nature', *Revue Francaise de Sociologie*, 34, pp. 495–524.

Lash, S. (1994), 'Expert-Systems or Situated Interpretation? Culture and Institutions in Disorganized Capitalism', in U. Beck et al., *Reflexive Modernization* (Cambridge: Polity).

Le Bris, S. (1993), *National Ethics Bodies* (Luxembourg: Council of Europe Press).

Leiss, W. (1972), *The Domination of Nature* (New York: George Braziller).

Lemke, J. L. (1985), 'Textual Politics: Heteroglossia, Discourse Analysis and Social Dynamics' (School of Education, Brooklyn College, City University of New York, mimeo).

Lemke, J. L. (1988), 'Discourses in Conflict: Heteroglossia, and Text Semantics', in J. D. Benson and W. S. Greaves (eds) *Systemic Perspectives on Discourse: Selected Papers from the Twelfth International Systemic Workshop* (New York: Ablex).

Lemonick, M. D. (1995), 'Seeds of Conflict', *Time*, 25 September, p. 78.

Lenk, H. (1992), *Zwischen Wissenschaft und Ethik* (Frankfurt: Suhrkamp).

Lex, M. (1995), 'Public Policy Issues Relating to Animal Biotechnology', in P. R. Wheale and R. McNally, (eds), *Animal Genetic Engineering: of Pigs, Oncomice and Men* (London: Pluto Press).

Lowe, Philip and Morrison, D. (1984), 'Bad News or Good News: Environmental Politics and the Mass Media', *Sociological Review*, 32, pp. 75–90.

Luhmann, N. (1981a), *The Differentiation of Society* (New York: Columbia University Press).

Luhmann, N. (1981b), 'Gerechtigkeit in den Rechtssystemen der modernen Gessellschaft', In N. Luhmann, *Ausdifferenzeirung des Rechts. Beitrage zur Rechtssoziologie und Rechtstheotie* (Frankfurt: Suhrkamp) (first published 1973).

Luhmann, N. (1986), *Ecological Communication* (Chicago: Chicago University Press).

Luhmann, N. (1993a), *Risk: a Sociological Theory* (translated by R. Barret) (New York: de Gruyter).

Luhmann, N. (1993b), *Das Recht der Gesellschaft* (Frankfurt: Suhrkamp).

Luhmann, N. (1995), *Social Systems* (Stanford CA: Stanford University Press) (original work published 1984).

Lukes, S. (1974), *Power: a Radical View* (London: Macmillan).

Lyotard, J. F. (1988), *The Différend: Phrases in Dispute* (translated by G. van den Abbeele) (Minneapolis: Minnesota Press) (original work published 1983).

Mariacher, A. (1991), 'Mehr Akzeptanz durch mehr Dialog', *Chemische Industrie,* 1, pp. 27–31.

McGee, M. C. (1980), 'The "Ideograph": a Link between Rhetoric and Ideology', *The Quarterly Journal of Speech,* 66.

McHale, J. (1992), *Medical Confidentiality and the Law* (London: Routledge).

McNally, R. (1994), 'Genetic Madness: the European Rabies Eradication Programme', *The Ecologist,* Vol. 24, No. 6, Nov/Dec, pp. 207–12.

McNally, R. (1995), 'Eugenics Here and Now', *The Genetic Engineer and Biotechnologist,* Vol. 15. 2 and 3, pp. 135–44.

McNally, R. (1996), 'Political Problems, Genetically Engineered Solutions: Sociotechnical Translations of Fox Rabies', in A. van Dommelen (ed.) *Coping with Deliberate Release: The Limits of Risk Assessment* (Tilburg and Buenos Aires: International Centre for Human and Public Affairs).

McNally, R. and Wheale, P. R. (1994), 'Environmental and Medical Biothetics in Late Modernity: Giddens, Genetic Engineering and the Post-Modern State', in R. Attfield and A. Belsey (eds) *Philosophy and the Natural Enviornment* (Cambridge: Cambridge University Press).

Meister, H. P. (1996), 'Community Advisory Panels in the U.S.A.', in H. Hermann (ed.) *Dialoge über Grenzen. Kommunikation bei Public–Private Partnership* (Köln: Carl Heymanns).

Melucci, A. (1980), 'The New Social Movements: a Theoretical Approach', *Social Science Information,* 19, pp. 199–226.

Melucci, A. (1985), 'The Symbolic Challenge of Contemporary Movements', *Social Research,* 52, pp. 789–816.

Melucci, A. (1989), *Nomads of the Present* (London: Hutchinson Radius).

Ministry of Housing, Netherlands (1994), 'Spatial Planning and the Environment: the Netherlands', in *Public Information and Participation in the Context of European Directives 90/219/EEC and 90/220/EEC* (The Hague).

Montgomery, J. (1991), 'Rights, Restraints and Pragmatism', *Modern Law Rewiew,* 54.

Morgan, D. (1992), 'What HFEA Did in its First Year', *Bulletin of Medical Ethics,* 17, p. 18.

Morgan, D. and Lee, R. G. (1991), *Human Fertilisation and Embryology Act 1990* (London: Blackstone).

Moscovici, S. (1968), *Essai sur l'histoire humaine de la nature* (Paris: Flammarion).

Moscovici, S. (1990), 'Questions for the Twenty-first Century', *Theory, Culture and Society*, 7, pp. 1–19.

Moser, I. (1995), 'Mobilising Critical Communities and Discourses on Modern Technology', in V. Shiva and I. Moser *Biopolitics* (eds) (London: Zed Books).

Munch, R. (1991), *Dialektik der Kommunikationsgesselschaft* (Frankfurt: Suhrkamp).

Munch, R. (1992), 'The Dynamics of Societal Communication', in P. Colomy (ed.) *The Dynamics of Social Systems* (London: Sage).

Munzer, S. (1993), 'An Uneasy Case against Property Rights in Body Parts', in *W. G. Hart Legal Workshop Paper* (London: Institute of Advanced Legal Studies).

Murphy, J. (1994), 'The Civil Liability of Obstetricians for Injuries Sustained during Childbirth', *Professional Negligence*, 10.

Murphy, R. (1995), 'Society as if Nature did not Matter', *British Journal of Sociology*, 46, 4, pp. 688–707.

Math, B. and Talay, I. (1996), 'Man, Science, Technology and Sustainable Development', in B. Nath, L. Hens and D. Devuyst *Textbook on Sustainable Development* (Brussels: VUB Press).

Nelkin, D. (1975), 'The Political Impact of Technical Expertise', *Social Studies of Science*, 5, pp. 35–54.

Nelkin, D. (1992), *Controversy* (London: Sage).

Nowotny, H. (1994), *Time: the Modern and Postmodern Experience* (Cambridge: Polity Press).

O'Donovan, K (1988), 'A Right to Know One's Parentage?', *International Journal of Law And the Family*, 27.

O'Mahony, P. (1991), 'Science and Industry', *Education and Training Technology International*, 28, 1.

O'Mahony, P. (ed.) (1996), *Final Report of the Project Sustainability and Institutional Innovation* (Cork: CESR, Unpublished MS).

O'Mahony, P. and Skillington, T. (1996), 'Sustainable Development as an Organising Principle for Discursive Democracy', in *Sustainable Development*, 4, 1, pp. 42–51.

Offe, C. (1992), 'Bindings, Shackles, Brakes', in A. Honneth et al. (eds) *Cultural-Political Interventions in the Unfinished Project of Enlightenment* (Cambridge, Mass: MIT).

Organisation for Economic Co-operation and Development (1988), *Biotechnology and the Changing Role of Government* (Paris: OECD).

Piatier, A. (1984), *Barriers to Innovation* (London: Frances Pinter).

Piore, M. and Sabel C. (1984), *The Second Industrial Divide: Possibilities for Prosperity* (New York: Basic Books).

Poferl, A. and Brand, K. W. (1995), 'Framing and Communicating Environmental Issues: the German Case' (Unpublished Report of Project No. 42, Co-ordinated by Eder, K. European University Institute, Florence).

Popper, K. (1974), *The Logic of Scientific Discovery* (London: Hutchinson).

Powell, W. W. and DiMaggio, P. J. (eds) (1991), *The New Institutionalism in Organizational Analysis* (Chicago: University of Chicago Press).

Putterman, D. M. (1994), 'Trade and the Biodiversity Convention', *Nature,* 371, 13 October, pp. 553–4.

RAFI (1994), *Conserving Indigenous Knowledge: Integrating Two Systems of Innovation* (New York: United Nations Development Programme).

Rehberg, K. S. 'Die "Offentlichkeit" der Institutionen: Grundbegriffliche Uberlegungen im Rahmen der Theorie und Analyse institutioneller Mechanismen', in G. Gohler (ed.) *Macht der Offentlichkeit: Offentlichkeit der Macht* (Baden-Baden: Nomos).

Roberts, T. (1996), 'Plant Patent Quagmire', *Nature,* 381, 20 June, p. 642.

Rudolph, E. (1989), 'The Role of Conjunctions and Particles for Text Connexity', in M. Conte, E., J. S. Petofi and E. Sozer (eds) *Text and Discourse Connectedness* (Amsterdam: John Benjamins Publishers).

Schomberg, R. von. (ed.) (1995), *Contested Technology: Ethics, Risk and Public Debate* (Tilburg, The Netherlands: International Centre for Human and Public Affairs).

Schomberg, von R. and Wheale, P. R. (1995), 'Human Genome Research', *Biotechnology and Development Monitor,* no. 25, December, pp. 8–11.

Schönefeld, L. (1996), 'Kommunikation mit der Nachbarschaft', in H. Herman (ed.) *Dialoge Über Grenzen. Kommunikation bei Public–Private Partnership* (Köln: Carl Heymanns).

Seltzer, M. (1992), *Bodies and Machines* (New York: Routledge).

Sessions, G. and Naess, A. (1991), 'The Basic Principles of Deep Ecology', in J. Davis (ed.) *The Earth First Reader* (Salt Lake City: Gibbs-Smith).

Simons, J. (1995), *Foucault and the Political* (London: Routledge).

Skillington, T. (1996), 'Illustrating the Reflection of Structure in Symbolic Action on the Environment: Reasserting the Old and Introducing the New through a Changing Policy Discourse' (Centre for European Social Research: University College, Cork, Working Paper).

Skillington, T. (1997), 'Politics and the Struggle to Define: a Discourse Analysis of the Framing Strategies of Competing Actors in a New Participatory Forum', *British Journal of Sociology,* 48, 3.

Smart, B. (1995), 'The Subject of Responsibility', *Philosophy and Social Criticism,* 21, 4, pp. 93–109.

Soper, K. (1995), *What is Nature?* (Oxford: Blackwell).

Spangenberg, J. (1993), 'Participation Overkill', *Die Tageszeitung, Berlin,* 10 April, p. 6.

Spolsky, E. (1994), 'Doubting Thomas and the Senses of Knowing', *Common Knowledge,* 3, 2, pp. 111–29.

Stehr, N. (1994), *Knowledge Societies* (London: Sage).

Stevenson, P. (1995), 'Patenting on Transgenic Animals: A Welfare/Rights Perspective', in P. R. Wheale and R. McNally (eds) *Animal Genetic Engineering: of Pigs, Oncomice and Men* (London: Pluto).

Street, J. (1992), *Politics and Technology* (London: Macmillan).

Strydom, P. (1987), 'Technology as a Collective Learning Process,' (Paper presented to the Annual Conference of the Sociological Association of Ireland, Dublin. 24–25 April).

Tarrow, S. (1989), *Democracy and Disorder: Protest and Politics in Italy, 1965–1975* (Oxford: Clarendon).

Tester, K. (1993), *The Life and Times of Post-Modernity* (London: Routledge).

Thibault, P. J. (1991), *Social Semiotics as Praxis: Social Meaning Making and Nabokov's Ada* (Minneapolis: University of Minnesota Press).

Thielemann, U. (1994), 'Unternehmerische Chemiepolitik zwischen Kooperation und Konfrontation: Sinn und Grenzen kooperativer Subpolitik', in Oikos (ed.), *Kooperationen fur die Umwelt Im Dialog zum Handelin* (Chur: Ruegger).

Thielemann, U. (1995), *Die Verantwortung der chemischen Industrie angesichts des bio- und gentechnologischen Fortschritts* (Institut für Wirtschaftsethik, St. Gallen, unpublished MS).

Thompson, M. (1991), 'Plural Rationalities: the Rudiments of a Practical Science of the Inchoate', in J. A. Hansen (ed.) *Environment Concerns. An Interdisciplinary Exercise* (London: Elvsevier).

Triseliotis, J. (1984), 'Obtaining Birth Certificates' in P. Bean (ed.) *Adoption: Essays in Social Policy, Law and Sociology* (London: Tavistock).

Turner, F. (1974), 'Rainfall, Plagues and the Prince of Wales', *Journal of British Studies,* 13, pp. 46–65.

Turner, F. (1978), 'The Victorian Conflict between Science and Religon', *Isis,* 69, pp. 356–76.

Ueberhorst, R. and de Man, R. (1992), *Zweite Frankfurter Studie zur Forderung chemiepolitischer Verstandigungsprozesse* (Frankfurt: Elmshorn).

Ulrich, P. (1977), *Die Großunternehmung als quasi-offentliche Institution: Eine politische Theorie der Unternehmung* (Stuttgart: C. E. Poeschel).

Ulrich, P. (1993), *Integrative Wirtschafts und Unternehmensethik – ein Rahmenkonzept* (Beitrage und Berichte des Instituts fur Wirtschaftsethik, St Gallen).

Van Dyck, J. (1995), *Manufacturing Babies and Public Consent* (London: Macmillan).

Van Peursen, C. A. (1970), *Strategie van de Cultuur* (Amsterdam: Elsevier).

Van Vliet, M. (1993), 'Environmental Regulation of Business: Options and Constraints for Communicative Governance', in J. Kooiman (ed) *Modern Governance* (London: Sage).

Velody, I. (1994), 'Constructing the Social', *History of the Human Sciences,* 7, pp. 81–5.

Verband der Chemischen Industrie ev (ed.) (1992), *Chemie im Dialog. Umweltleitlinein* (Frankfurt: Verband der Chemischen Industrie).

Vidal, J. and Carvel, J. (1994), 'Lambs to the Gene Market', *The Guardian,* 25 November, p. 25.

Virilio, P. (1986), *Speed and Politics: An Essay on Dromology* (translated by Mark Polizzotti) (New York: Semiotexte).

Volosinov, V. N. (1973), *Marxism and the Philosophy of Language* (translated by L. Matejka and I. R. Titunik) (New York: Seminar).

Warnock Committee. (1984), *The Warnock Report* (HMSO, London).

Warnock, M. (1985), *A Question of Life* (Oxford: Blackwell).

Webster, A. (1990), 'The Incorporation of Biotechnology into Plant-breeding in Cambridge', in I. Varcoe et al. (eds) *Deciphering Science and Technology* (Basingstoke: Macmillan).

Weingart, P. (1978), 'The Relation between Science and Technology: a Sociological Explanation' in W. Krohn, Layton and P. Weingart (eds) *The Dynamics of Science and Technology: The Sociology of the Sciences, Volume 2* (Dordrecht: D. Reidel).

Weise, E. (1988), 'Oko-soziale Marktwirtschaft statt "neuer" Chemiepolotik', in M. Held (ed.) *Chemiepolitik: Gesprach über eine neue Kontroverse. Beitrage und Ergebnisse eine Tagung der Evangelischen Akademie Tutzing: 4 bis 6 Mai 1987,* pp. 60–7 (Weinheim: VCH Verlagsgesllschaft mbH).

Welsch, W. (1996), 'Aestheticisation Processes: Phenomena, Distinctions and Prospects', *Theory, Culture and Society,* 13, 1.

Wessels, U. (1994), 'Genetic Engineering and Ethics in Germany', in A. Dyson and J. Harris (eds) *Ethics and Biotechnology* (London: Routledge).

Wheale P. R. and McNally, R. (1986), 'Patent Trend Analysis: the Case of Genetic Engineering', *Futures,* October, pp. 638–57.

Wheale, P. R. and McNally, R. (1988a), *Genetic Engineering: Catastrophe or Utopia?* (Hemel Hempstead: Wheatsheaf).

Wheale P. R. and McNally, R. (1988b), 'Technology Assessment of a Gene Therapy', *Project Appraisal,* 3, 4, December, pp. 199–204.

Wheale, P. R. and McNally, R. (eds) (1990), *The Bio-Revolution: Cornucopia or Pandora's Box?* (London: Pluto Press).

Wheale, P. R. and McNally, R. (1993), 'Biotechnology Policy in Europe: a Critical Evaluation', *Science and Public Policy,* 20, 4, August, pp. 261–79.

Wheale, P. R. and McNally, R. (eds) (1995), *Animal Genetic Engineering: Of Pigs, Oncomice and Men* (London: Pluto Press).

Wheale, P. R. and McNally, R. (1994), 'What Bugs Genetic Engineers about Bioethics', in A. Dyson and J. Harris (eds) *Ethics and Biotechnology* (London: Routledge).

Wheale, P. R. and McNally, R. (1996), 'On How People Can Become "The Prince": Machiavellian Advice to NGOs on GMOs', in A. van Dommelen (ed.) *Coping with Deliberate Release: the Limits of Risk Assessment* (Tilburg and Buenos Aires: International Centre for Human and Public Affairs).

Whitbeck, C. (1991), 'Ethical Issues Raised by the New Medical Technologies', in J. Rodin and A. Collins (eds), *Women and New Reproductive Technologies* (New Jersey: Lawrence Erlbaum).

White, S. (1991), *Political Theory and Postmodernism* (Cambridge: Cambridge University Press).

White, S. (ed.) (1995), *The Cambridge Companion to Habermas* (Cambridge: Cambridge University Press).

Wiedmann, K. P. and Ries, K. W. (1992), *Risikokommunikation und Marketing,* Arbeiten zur Risiko-Kommunikation, No. 29, Julich: Programmgruppe Mensch, Umwelt, Technik (MUT) der KFA Julich GmbH.

Williams, Rhys H. (1995), 'Constructing the Public Good: Movements and Cultural Resources', *Social Problems,* 42, 1.

Willke, H. (1983), *Enzauberung des Staates: Überlegungen zu einer sozietalen Steuerungstheorie* (Konigstein: Athenaum).

Willke, H. (1992), *Ironie des Staates: Grundlinien einer Staatstheorie polyzentrischer Gesellschaft* (Frankfurt: Suhrkamp).

Zolo, D. (1992), *Democracy and Complexity: A Realist Approach* (Cambridge: Polity).

Index

Index